Specialist Control

*The Publications Committee
of the Académie Royale des
Sciences (Paris)
1700–1793*

Figure 1: Commemorative medal struck for the Académie Royale des Sciences. Undated eighteenth-century restrike in bronze. Obverse signed by F. Mauer, fils with bust of Louis XIV and legend, "LUDOCIVUS XIIII REX CHRISTIANISSIMUS" — "The Most Christian King Louis XIV." Reverse: Minerva seated facing left amid scientific instruments and botanical specimens with legend, "NATURÆ INVESTIGANDÆ ET PERFIC. ARTIB." — "Nature is to be investigated and the arts perfected." Exergue: "REGIA SCIENTIARUM ACADEMIA INST. M.DC.LXVI" — "The Royal Academy of Sciences founded in 1666."

TRANSACTIONS
of the
AMERICAN PHILOSOPHICAL SOCIETY
Held at Philadelphia
For Promoting Useful Knowledge
Volume 92, Part 3

Specialist Control

☙❧

The Publications Committee of the Académie Royale des Sciences (Paris) 1700–1793

James E. McClellan III

American Philosophical Society
Philadelphia • 2003

ISBN: 0-87169-933-8
US ISSN: 0065-9746

Library of Contress Cataloging-in-Publication Data

McClellan, James E., 1946–
    Specialist control : the Publications Committee of the Académie royale des sciences
(Paris), 1700–1793 / James E. McClellan, III.
        p. cm. — (Transactions of the American Philosophical Society ; v. 93, pt. 3)
    Includes bibliographical references and index.
    ISBN 0-87169-933-8 (pbk.)
        1. Académie royale des sciences (France) Comité de librairie—History. 2.
Science—France—History—18th century. I. Title. II. Series.

Q46.A15M35 2003
509.44'09'033—dc21

                                                                2003048159

Contents

Abbreviations

AdS: Institut de France, Académie des Sciences, Service des Archives, 26, quai de Conti, 75005 Paris.

AdS, DG 31: AdS, "Dossier Général, Carton 31." This is a substantial collection of documents that pertain to the institutional history of the Academy of Sciences.

APS Fougeroux papers: American Philosophical Society Library, Philadelphia, Manuscript Collections, Fougeroux de Bondaroy Papers, B/F8245, "Séances de l'Académie Royale des Sciences, 1786–1789."

HARS: The *Histoire* section of the Academy's annual volume of *Histoire et Mémoires de l'Académie Royale des Sciences.* See also **MARS**. Regarding the nominally annual volume of the *Histoire et Mémoires*, note that the *Histoire* section summarized the work of the Academy a particular year, and the *Mémoires* section presented the formal papers of academicians. References to the *Histoire et Mémoires* are to the year the volume covered, not the year of its publication.

HBI: Hunt Botanical Institute, Carnegie-Mellon University, Pittsburgh, PA.

Lalande, "Collection": Jérôme de Lalande ms, "Collection de ses Règlemens et Déliberations;" ms. in the Biblioteca Medicea-Laurenziana, Florence, cited in Hahn, *Anatomy*, xiv. I thank my colleague, Arnold B. Urken, for providing me a copy of this rich and useful document. Photographs of parts of this manuscript are in AdS, DG 31.

MARS: The *Mémoires* section of the Academy's annual volume of *Histoire et Mémoires de l'Académie Royale des Sciences.* See also **HARS**.

RCL: Ms. "Registres des Déliberations du Comité de Librairie," AdS. There are two volumes of manuscript registers of the Comité de Librairie. RCL volume I (189 ms. pp.) covers the period from January 1749 through December 1769; RCL volume II (77 ms. pp.) covers the period from January 1770 through June 1780. Citations (in the form RCL I/x [date]; RCL II/x [date]) refer to register volume, page number, and date. The date refers to the month papers were read to the Academy and so inscribed on the books of the Comité de Librairie, not the date on which Comité met to consider papers. RCL II is paginated only to page 71; pages [72] to [77] are indicated in brackets. A separate folio page in Condorcet's hand recording the meeting of June 23, 1780 concludes the second volume of the registers. Figure 4 below shows the two bound volumes in question.

RCT: Registres du Comité de Trésorerie, AdS.

PS: Mss. "Pochettes de Séance," AdS.

PV: Mss. "Procès verbaux des séances," AdS.

SE: The *Savants Étrangers* series of the Académie Royale des Sciences.

List of Illustrations

Figures 2 to 6 are reproduced by permission from originals at the Académie des Sciences of the Institut de France, Paris. Figures 1, 7, and 8 are of items from the author's collection. All figures have been electronically enhanced.

Preface

I have come to criticize fellow historians and myself for not regularly acknowledging first and foremost that their own writings are historical products. I now hold that every piece of history writing should incorporate an account of how it came to be. True to my convictions, I report that this study began abruptly in the mid-afternoon of Wednesday, April 5, 1995 with an unexpected question.

The scene was Daniel Roche's seminar at the École Normale Supérieure in Paris. I was presenting the results of a study I had made of the scientific papers published in the *Mémoires* of the French *Académie Royale des Sciences*. Toward the end of the seminar, Roche pointedly asked how the papers came to be published, that is, through what editorial processes did they appear in print. I mumbled something about the requirement for papers to be read before the Academy, but nobody present—myself especially included—had an answer to Roche's question of how exactly the Academy chose the papers it published in its *Mémoires* and other scientific series. The next day I hurried over to the Service des Archives of the Académie des Sciences, and I immediately fell upon the detailed registers of the Academy's Comité de Librairie, its publications committee (Fig. 2). (The registers are stored in plain sight on a bottom shelf on the left just as one enters the present archival storage room.) Historians of the Academy have always known about the Comité de Librairie and had logged the existence of its registers, but no one had studied them in detail. I set about to do so in order to answer Roche's seemingly simple question.

The investigation and, now, this presentation unfolded in several stages since that afternoon on the rue d'Ulm, and along the way the author and his project have been aided in many ways and by many individuals he is pleased to acknowledge. Most of the research for this study was done at the Académie des Sciences in Paris. An intensive reading of the extant records began there the day after Roche's seminar, and work on this project continued at the Académie until just recently. I am again deeply indebted to the staff of the Service des Archives of the Académie des Sciences for their help and friendship, notably Mme Claudine Pouret, M. Le Roy, Mme Mine, and previously Mme Christiane Demeulenaere-Douyère and now Mme Florence Greffe as head of that service.

A sabbatical leave in 1999 and 2000 from my home institution of Stevens Institute of Technology and a complementary sabbatical fellow-

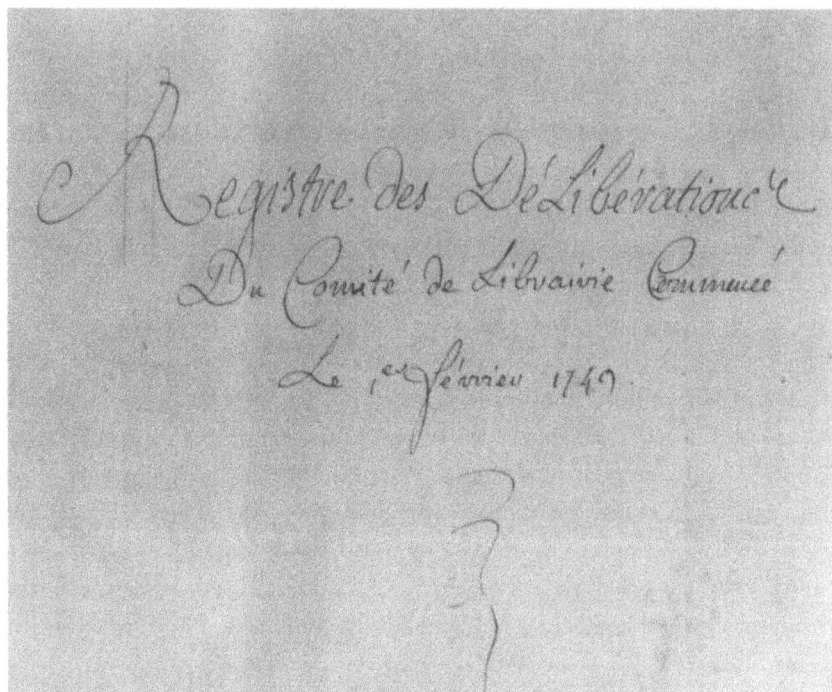

Figure 2: Inscribed leaf from the minute book of the Comité de Librairie, RCL I.

ship from the American Philosophical Society (APS) materially and morally supported this project. By 1999, an initial written version had been read and commented upon by other scholars, so the sabbatical leave from Stevens and the fellowship from the APS allowed me to pursue follow-up research along specific lines. During this period I spent a number of days over several weeks working at the Library of the American Philosophical Society in Philadelphia. At the APS I have particularly to thank Mr. Roy Goodman and Ms. Eleanor Roach for their assistance and good will. In the later stages of this project it was a pleasure to work with Ms. Mary C. McDonald, editor of APS publications, and I thank her for her professionalism and for being so nice.

In 1999 and 2000, in addition to return trips to Paris where I was able to pursue the archival trail, I also worked in London at the Royal Society of London and in Pittsburgh at the Hunt Botanical Institute. The staff at both institutions were gracious professionals aiding a wandering scholar, but I cannot omit specific mention of Ms. Angela Todd at the Hunt who was so helpful in arranging my visit and assisting me afterward.

This monograph has gone through an unusual number of iterations to arrive at its present form. The initial version has been mentioned.

Professors Charles C. Gillispie and Alice Stroup were kind enough to read it and to offer detailed remarks, and I thank them for their useful suggestions and for restraining the author's enthusiasm over all the juicy bits he had found in the minute books of the Comité de Librairie. This same initial version was submitted as a journal article and was refereed by three anonymous readers who, over two rounds of comments, signaled the article's shortcomings and proposed avenues for expanded consideration. (The irony of a paper about refereeing being refereed did not escape general notice.)

Professors Alice Stroup and David J. Sturdy read the penultimate version, and their learned and insightful readings teased out more than one analytical point that deserved clearer treatment. Incorporating their recommendations has raised the quality of this work considerably. I look forward to standing Alice and David dinner in Paris as a token of my gratitude. I likewise thank my colleague and friend, Harold Dorn, who burnished a late draft and whose sense of style effected further improvements. Thanks also go to Mme Marie-Jeanne Tits-Dieuaide who graciously contributed relevant findings from her own archival research, to M. Gérard Barré of the Maison Platt in Paris for helping identify the bronze medal depicted in Figure 1, and to Dr. Murray C. McClellan for his help with Latin expressions.

I gave four presentations that connect to the present study in various ways, and I am grateful to the organizers and audiences involved. I had the pleasure of participating in the 1998 Princeton Workshop in History of Science with a paper entitled "Science and Colonialism in the Old Regime." That exercise sharpened my thinking about transmission of knowledge across cultures, a theme that briefly, but crucially appears at the end of Chapter 7. In 1999 I was fortunate to attend the international conference, "Règlement, usages et science dans la France de l'Absolutisme," (Paris, 8–10 June 1999). There, I gave a first-cut screening of what I had learned of the Comité de Librairie, and the contributions of the audience and the intense experience of the conference clarified many details and pushed the study in several useful directions.[1] In the spring of 2001, relying heavily on material extracted from the records of the Comité, I twice spoke about scientific norms in historical context. The first was a public lecture at the New York Academy of Sciences; the second was a presentation at yet another Paris seminar, Éric Brian's, whom I thank for his invitation to present some of this material. The critical responses on both occasions highlighted many issues, and the results have been folded into this fuller account.

The article version did not work out, and as the editorial constraints of an article relaxed, the present monograph format fell naturally into place. This study of the Comité de Librairie of the old Académie Royale des Sciences touches on larger issues, but its own mission is more modest: to present a comprehensive and authoritative history of the Academy's Comité de Librairie. The hope is that this study will be useful to those who come across it. Of course, the author assumes full responsibility for all errors and shortcomings.

As already noted, the American Philosophical Society in Philadelphia played a large role in the coming-into-being of the present work. The author owes an additional debt of gratitude to the APS for judging this work suitable for its *Transactions*. This distinguished imprint of the APS seems a more than fitting publication outlet for this piece about scientific publishing.

Note

1. These initial perspectives are presented in McClellan, "Bilan Public et Processus Privés."

ᎦᏬ Chapter One ᏫᎧ

Historiographical Contexts

The monograph at hand documents the history of the Publications Committee or *Comité de Librairie* of the French *Académie Royale des Sciences* during the period of the Committee's existence from 1700 to 1793. Put another way, the present work is an in-depth scholarly treatment of one administrative committee in one scientific institution over the course of roughly one century. One might reasonably ask, why bother? Several substantial reasons explain why the Paris Academy's Comité de Librairie commands interest and why one would bring its story to public attention.

The scientific and historiographical importance of the Académie Royale des Sciences (Paris, 1666–1793) constitutes an underlying rationale for investigating its Comité de Librairie. The French Académie Royale des Sciences was the leading scientific academy and center of natural science in the eighteenth century. Among the reasons for the scientific and scholarly reputation of the institution are the ninety-three volumes of the Academy's famous *Histoire et Mémoires* (1699–1790), the foremost scientific series of the century and a series controlled and published by the Academy's Comité de Librairie (Fig. 3). Not surprisingly, the Académie Royale des Sciences of Old-Regime France is well studied and comes with an historiographically well-developed literature.[1] By bringing into focus the heretofore unstudied Comité de Librairie, the present piece hopes to add a new and useful perspective on the workings of the Academy and of French science in the Old Regime.

It should be emphasized that the Comité de Librairie was a powerful standing committee of the Paris Academy of Sciences. It was one of only two standing committees along with the Comité de Trésorerie, the latter formally established only in 1725. The Comité de Librairie was powerful because its ten members controlled the substantial published output of the

Figure 3: Frontispiece and title page from the inaugural 1699 volume of *Histoire et Mémoires de l'Académie Royale des Sciences* published in 1702. Frontispiece depicts Minerva seated with portrait of Louis XIV and surrounded by the insignia of various scientific disciplines. Note the *Observatoire royal* in the background.

Academy from 1700 to 1793, including the above-mentioned *Histoire et Mémoires* series and travel supplements, the *Savants Étrangers* series, volumes of winning prize essays, and indirectly the *Description des Arts et Métiers* and the *Connaissance des Temps*, collectively the largest body of scientific and technical work published anywhere in the world in the eighteenth century.[2] Furthermore, although scholars have known of the Comité de Librairie and its archival residue, no one to date has systematically investigated the group or its records.[3] The power of the Comité in the world of contemporary French science and the fact that the Comité has not been studied until now are certainly encouragements to do so.

A secondary, but still valuable historiographical connection embeds the Academy's Comité de Librairie more directly in the history of the scientific press. As outlined in Chapter 2, the control exercised by the Comité de Librairie represents a landmark in the history of the scientific press, and that finding justifies delving into the Comité's history for what the details add to the not insignificant literature surrounding the history of scientific publishing.[4]

The most compelling reason for presenting this study, however, lies in the fact that with the Academy's Comité de Librairie, scientists themselves—the producers of knowledge—for the first time gained control

over the process of publishing the results of their research. With the
Comité de Librairie a novel type and level of professional autonomy
appeared in the history of science, in which scientific institutions and the
producers of science first came to oversee the publication of science and
making knowledge public. The individuals involved were savants, of
course, and academicians, not scientists in a nineteenth-century sense,
but more than any previous scientific group the Paris Academy and its
Comité de Librairie came to oversee scientific publishing and to shape the
content of what appeared as natural science.

It is beyond the scope of this study to treat in any substantial way the
history of mechanisms for social and institutional control of knowledge
and knowledge-making prior to the Comité de Librairie, but a few remarks
along these lines may highlight the significance of the novelties represented
by the Comité de Librairie. From the Middle Ages forward, systems of
"private knowledge" in some measure circulated *sub rosa*, and there was
little formal control over the content of alchemy, numerology, botanical and
medical secrets, witchcraft of various shades, and other heterodox points of
view.[5] The examples of a Paracelsus, the occasional court alchemist, or the
prisca sapentia dear to Newton and others evidence this undercurrent of
secret knowledge systems and traditions that continued through the seventeenth century. By the same token, knowledge in the public realm seems
always to have been subject to vigilance by some public authority, notably
in the West by the Roman Catholic Church. In this connection one can
point to the submission of Roger Bacon to Franciscan authorities in the
thirteenth century, to the suppression of Giambattista della Porta's *Magia
naturalis* (1558) and his *Accademia Secretorum Naturae* by the Inquisition
around 1560, the trial and execution of Giordano Bruno in 1600, not to
mention Galileo's fate at the hands of Church authorities. A free market-
place of ideas did not exist for those who would inquire into nature and try
to make their findings public. Descartes suppressing his *Le Monde* on learn-
ing of Galileo's travails comes to mind in this regard. The only question
concerned who was the controlling authority.

Circumstances began to change in the seventeenth century. The
publication of Galileo's *Assayer* by the *Accademia dei Lincei* in 1623
heralded the potent new role that scientific societies would play in scientific publishing and in specialist control over the scientific press.[6] But in
seventeenth-century Italy at least the *Assayer* and other works of science
still had to receive an imprimatur from Church censors. Matters developed further with the advent of the *Philosophical Transactions* of the Royal
Society of London (1666) and the early publications of the Académie

Royale des Sciences. The institutional or semi-institutional basis for these and other contemporary scientific publications makes the point that, beginning in the 1660s, the natural philosophical enterprise achieved a new stature and organizational form and that the world of scientific discourse had narrowed somewhat to more restricted communities.[7] The producers of knowledge were becoming their own publishers.

Changing concepts and practices surrounding the publication of science in the seventeenth century can be usefully highlighted in this connection.[8] Consider Newton's paper on optics published in 1672 in the *Philosophical Transactions* of the Royal Society of London. Beginning with the salutation, "Dear Sir," it was nominally a letter written to the Secretary of the Royal Society, Henry Oldenburg. Yet, appearing in a journal sanctioned by the Royal Society, it was simultaneously an early scientific paper of a type that continues to be the mainstay of scientific publication today. As a transition between the personal letter and the published formal paper, the example testifies to the social and institutional fabrication of the scientific paper or what Simon Shaffer has termed the advent of "literary technologies." The historical development of a set of modern mechanisms surrounding the publication of scientific papers appeared in the historical setting examined here, and it is noteworthy in this regard that Shaffer pinpoints the setting where "literary technologies" first appeared amid "the academicians and naturalists of the Enlightenment."[9]

Indeed, it was with the Parisian Academy of Sciences and its Comité de Librairie that scientists themselves—the makers of new knowledge about nature—for the first time fully controlled the process of seeing material into print and thereby to the public at large. In contrast to the Royal Society of London and some other contemporary academies, the Academy and Comité exercised a formal power to "examine and judge" the scientific contributions of members and what appeared in print in the annual volume of *Histoire et Mémoires* and the Academy's other scientific publications. (Chapter 2 elaborates this point.) In other words, the Paris Academy through its Comité de Librairie was unique in being the guarantor of the accuracy and reliability—in short, the truth of the claims to knowledge it published and sanctioned. The Academy and the Comité thus exercised unprecedented authority over the scientific enterprise and the production and dissemination of knowledge. Not surprisingly, the Comité became a ferocious gatekeeper.[10]

A related topic informing this narrative concerns the question of social norms in the natural sciences and the historical development of normative behaviors and practices in scientific communities.[11] A certain wariness is called for in thinking about scientific norms because today's norms are plainly not normative but have evolved under changing historical circumstances. That said, however, the period and episode under investigation here were pivotal in the historical and social development of modern social norms displayed in scientific practice. Prior to the scientific societies of the seventeenth century, to pick one example, priorities and discoveries were sometimes protected using anagrams, and one recalls that Kepler famously misconstrued Galileo's anagram regarding the phases of Venus. In contrast, scientific priority came to be protected as a matter of course by the Paris Academy of Sciences and the Comité de Librairie. Similarly, prior to the Academy and the Comité de Librairie the author is not aware of anything like the refereeing and peer review processes that came to develop under their auspices.

The next point takes off from the observation that in recent years historians have problematized science as an object of study. That is, many historians have come to question the notion of science, especially in the early-modern period, as a unitary enterprise amenable to straightforward historical analysis.[12] They are right to have undertaken this reappraisal, not least as an antidote to earlier views that accepted science uncritically as a single entity whose content might change over time but whose goals and essence remained the same. Rejecting "science" as an unproblematical starting point for inquiry, this new historical trend posits a multiplicity of possibilities and nuances, grounded in precisely situated historical contexts, for what constituted inquiry into nature at different times under changing circumstances by various communities of practitioners.

Much argues for the reasonableness of this approach to thinking about science in and around the Academy in the eighteenth century. Condorcet, for example, held a view of mathematical science much at odds with the life sciences of his day.[13] Similarly, Lavoisier's new chemistry of the 1770s and 1780s differed not only in content, but *qua* science from what passed for the same in the Academy earlier in the eighteenth century. To an extent, these considerations inform the narrative that follows, as, for example, in examining the attitude of the Comité de Librairie at mid-century toward the antiquated medical views of associate academician P.-J. Malouin.[14] Also, in what follows an attempt has been made to use expressions like "natural knowledge" and the "makers of knowledge" rather than "science" and "scientists," as more appropriate for the eighteenth century.[15] By the same token, because the present study focuses on an institution (the Academy of Science) and on the functioning of a bureaucratic entity (the Comité de Librairie) a concern to rigorously categorize or contextualize the nature of "science" in every instance that comes up would seem to be less imperative. For present purposes it suffices to note that science was whatever the Academy and Comité said was science.

Yet another consideration proved relevant to examining the Comité de Librairie: the social process that transforms what is initially local or private knowledge, as in first making a discovery, into public knowledge as might appear in public media or accepted in scientific circles.[16] Put another way, this process of publication (in the sense of making public) carries novelty from a private to the public domain. The Comité de Librairie presents itself strongly in this regard because publication per se represents a key juncture—indeed, a bottleneck—in that longer pipeline of knowledge-making. Before the Comité de Librairie in Paris, knowledge made its way among savants and the public in a variety of ways: privately or semiprivately through correspondence, through social and scientific networks of various sorts, and through printers, booksellers, and the circulation of printed material. In this connection one thinks, for example, of the correspondence of Henry Oldenburg—all the more so because of his position as Secretary to the Royal Society of London—or perhaps of Tycho Brahe and his Uraniborg and Sterneborg, or the publication and circulation of Copernicus' *De revolutionibus orbium coelestium*. The point, however, is that prior to the eighteenth century and the creation of the Comité de Librairie, nowhere in this general process of bringing new knowledge to the public did the makers of knowledge have a collective hand in, much less control over, what their peers or less reputable others might see into print. In terms of the trajectory of knowledge-making suggested here, the historical importance of the Comité de Librairie derives from the fact that it inserted a new filter for screening and sifting the scientific production of others, and for the first time specialists gained veto power over published science and the making public of knowledge more generally. That was plainly a nontrivial step in the social and institutional evolution of modern science, one worthy of a detailed case study.

A last point may be alluded to briefly, and that concerns who controls knowledge of nature and what that power gives them. A variety of accounts describe and explain the natural world around us, and "local" knowledge (of a peasantry, say) has always floated alongside more refined communities of learned discourse. In the eighteenth century, however, an official, sanctioned science came into being to assert itself in what might be seen as a sort of intellectual imperialism. With the Academy in general and the Comité de Librairie in particular new mechanisms manifested themselves for legitimating knowledge, for monitoring claims about nature, and for imposing sanctioned scientific authority over local or competing knowledge systems or claims. Examples of the exercise of the new, professional power vested in the Academy and in the Comité range from the most mundane decisions, to Mesmer's explusion from the community of scientists, or to the translation and assimilation of non-Western systems of knowledge the French encountered in their developing colonial empire. In short, with the Comité de Librairie scientific experts took command, and for the first time the producers of knowledge decided—or at least a coterie of scientific specialists decided on their behalf—what the public was legitimately supposed to think about nature. In so doing the institutionalized scientific community (the Academy's Comité de Librairie in this case) legitimated its own authority and a sovereign right to decide in matters concerning nature or the sciences.[17] That authority did not go uncontested.

A rich gamut of resources, mostly archival, enable a fine-grained study of the Comité de Librairie. That material will be encountered *en route*, but worthy of separate mention here are the detailed, extant registers of the meetings of the Comité de Librairie spanning more than three decades in the heart of the Old Regime, from 1749 to 1780 (Fig. 4). The two bound volumes of registers record monthly committee meetings for those

Figure 4: The two surviving bound minute books of the Comité de Librairie, RCL I and RCL II.

years, and they virtually beg to be examined.[18] Other documents allow one to chart the story of the Comité de Librairie over the long arc from its formal inception in 1700 through its demise along with the rest of the Academy in 1793. These precious resources offer a vivid panorama of this key scene in the historical development of the scientific enterprise. They reveal what the Academy published and why, how the Comité helped establish norms, how it monitored priority claims and adjudicated scientific disputes, how it guarded access to the press, and how, indeed, it played a role in shaping the production of scientific knowledge. The Comité's records also show its role as publisher and its dealings with the contemporary book and printing trades, the practical *sine qua non* of producing volumes of science. The study at hand presents findings regarding the statutory basis of the Comité and its membership structure, wherein it emerges that power accrued to a narrow group of academicians. The story of the Comité de Librairie is absorbing in its own right, and what we learn adds to our understanding of the Academy of Sciences, the development of specialist control over the scientific press, and the history of science in the eighteenth century generally.

Notes

1. For an entrée into this literature, see Brian and Demeulenaere-Douyère, eds., *Histoire et mémoire*; McClellan, "Académie des Sciences." Hahn's *Anatomy* remains a landmark in the literature. See also Stroup, *A Company of Scientists* and her *Royal Funding*; Sturdy, *Science and Social Status*; Brian, "L'Académie royale des science;" Paul, *Science and Immortality*; Gillispie, *Science and Polity*, 81–99.

2. Halleux et al., *Les publications de l'Académie* documents and provides bibliographic and research access to all the volumes (and their content) published by the Academy in the period prior to the French Revolution. The author's "A Statistical Overview" is relevant in this connection. Guénoun, "Les publications," is an original treatment also to be consulted; see further Gillispie, *Science and Polity*, 337–368.

3. Regarding the Comité de Librairie, see Brian and Demeulenaere-Doyère, eds., *Histoire et mémoire*, 75–76; Crépel, "Une curieuse lettre," does make exemplary scholarly use of the registers of the Comité for one particular case.

4. For an entrée into this rich literature and for background concerning the contemporary scientific press, consult McClellan, "Scientific Journals," and Sgard's monumental *Dictionnaire des journaux* and his *Dictionnaire des journalistes*. Gascoigne, *A Historical Catalogue*; Meadows, ed., *Development of Scientific Publishing*, and Manten, "Development of European Scientific Journalism" are likewise important sources. See also Hunter, ed., *Thornton and Tully*; Ford, "Eighteenth-Century Scientific

Publishing;" and Vittu, "Périodiques." Still valuable are Thornton and Tully, *Scientific Books, Libraries and Collectors*; Kronick, *A History of Scientific and Technical Periodicals*, and his *Scientific and Technical Periodicals of the Seventeenth and Eighteenth Centuries: A Guide*.

5. On these traditions, see Eamon, *Science and the Secrets of Nature* and Smith, *The Business of Alchemy*. For a sense of the developed historiography surrounding these topics, consult Cohen, *The Scientific Revolution*, esp. chapters 3.3 and 3.4.

6. For background, see McClellan, "Scientific Institutions," and the same author's earlier *Science Reorganized*, esp. chapt. 2; Meadows, *Communication in Science*, 66–90 and his *Communicating Research*, 5–11. In examining habits of early-modern erudition, Grafton, *The Footnote*, chapters 5 and 6, provides an insightful contrast to these institutional developments.

7. On the latter point, see the seminal work by Shapin and Schaffer, *The Leviathan and the Air Pump*. Note that, as discussed in the next chapter, the *Philosophical Transactions* did not become an official publication of the Royal Society until 1752.

8. Here one can only allude to a sophisticated and growing literature probing linguistic and rhetorical developments in science that parallel the institutional and journalistic ones of concern here. See especially the volumes edited by Lenoir, *Inscribing Science* and *Instituting Science*, both part of the "Writing Science" series published by Stanford University Press. Bazerman, *Shaping Written Knowledge* offers a suggestive, postmodern analysis of the development of the scientific paper and rhetorical strategies involved in the new literary form; his Part Two: "The Emergence of Literary and Social Forms in Early Modern Science" focuses on the *Philosophical Transactions* through the eighteenth century, on article publishing, and on the impact of the new form of the scientific paper on scientific communities. In their highly technical volume Halliday and Martin, *Writing Science*, examine linguistic and grammatical factors that readers may wish to consider in thinking about science in the period under consideration here. The same can be suggested for the recondite work by Gumbrecht and Pfeiffer, *Materialities of Communication*.

9. Shaffer, "The Leviathan of Parsonstown," 182; see also related paper by Daston, "The Language of Strange Facts."

10. Meadows, *Communication in Science*, 68, is no doubt right to suggest that possessing an independent imprimatur made the Academy and other similarly endowed early learned societies especially cautious in what they published for fear of compromising their privilege.

11. Mulkay's 1976 "Norms and Ideology" is a still useful analysis which presents "norms" as "vocabularies of justification." The present study seeks to extend Mulkay's work with the notion that norms in science are not normative, but historical. In 1971 Roger Hahn was already aware of the role of the Academy in constructing

normative behavior in the world of science; see Hahn, *Anatomy*, x. For references to the standard literature on norms in science, see Meadows, *Communication in Science*, p. 35.

12. Steve Shapin gives voice to this view in his *Scientific Revolution*, Introduction.

13. Baker, Condorcet, 42–43.

14. See below, chapter 7.

15. Gender-neutral language is also used here throughout, avoiding such otherwise useful expressions as "men of science" for "scientist." It should not need pointing out, however, that with the exception of a few women artists encountered below in chapter 8, only men figure in the story of the Comité de Librairie.

16. See sources cited in note 8 and Stewart, *The Rise of Public Science*, and Golinski, *Science as Public Culture*. Jacobi, *Diffusion et vulgarisation* and his *La communication scientifique*, illuminates another dimension of this same process, the complicated and nonobvious ways in which materials originally intended for communities of scientific peers are popularized and diffused to larger, nonexpert audiences. On these points, see also Meadows, *Communication in Science*, pp. 206–230.

17. The resulting tension between scientific authority and public resentment is a guiding theme in Hahn, *Anatomy*; in this connection see also Darnton, *Mesmer and the End of the Enlightenment*, and Conclusions, below.

18. These records are described more fully above in Abbreviations (page vii).

๑ Chapter Two ๑

The Academy's Publication Privilege and

the Powers of the Comité de Librairie

The Académie Royale des Sciences was founded in 1666. Louis XIV granted his Academy the authority to govern itself and to manage its own publications. In the first years of its existence the Academy was known for its communal, Baconian approach to producing knowledge, and so issues of assigning credit for scientific work or certifying priority or accuracy did not stand at the forefront of the institutional agenda.[1] By degrees, however, the Academy's initial collectivist approach gave way to a new pattern of individual academicians writing and presenting accounts of their own research. As this transition took place the Academy evolved into an independent arbiter and the supreme judge of science, exercising the power to examine and certify scientific work presented by its members and outsiders. In this way the Academy reoriented itself from a collective producer of knowledge to a collective judge and so established procedures it maintained through the eighteenth century.

From the beginning the Academy possessed its own imprimatur, and in the period from 1666 until its reform in 1699, the Academy published eight major volumes of memoirs and a number of other works.[2] It also came to lend its imprimatur for outside publication. How exactly the academicians chose what appeared in print in the Academy's seventeenth-century phase remains somewhat obscure. Presumably concensus determined the content of the early works issued under the Academy's collective authorship, such as its *Mémoires pour Servir à l'Histoire Naturelle des Animaux* (1671). On the other hand individual authors are acknowledged in the *Recueil de plusieurs traitez de Mathématiques* published in 1676, and in time meetings of the Academy shifted to refereeing the scientific production of individual savants.[3] By 1678 a committee system had emerged to evaluate projects submitted to the Academy; by 1685 the Academy was issuing standardized "approvals," and in 1688 the Academy formally adopted the rule that academicians wishing to publish materials presented at its meetings first had to have them reviewed by the Academy.[4] In this way a pattern of editorial and scientific control emerged within the seventeenth-century Academy that became incorporated in the eighteenth-century Comité de Librairie.

The Comité de Librairie came into official existence in 1700 attendant to the organizational reform of the Academy of 1699.[5] The formal Letters Patent of 1699 did not mention the Comité de Librairie, but they granted the Academy the right to publish its own volumes, and they required that the Academy approve papers submitted by academicians for publication in its volumes and for outside publication, if authors wanted to use the title of Academician.[6]

The Academy thus possessed a powerful formal privilege to examine the work of academicians and to publish independently of the elaborate state apparatus then in place for censoring books. The king's council overturned the Academy's original privilege because it was cast in perpetuity, and another was prepared in 1717.[7] In it, Louis XV granted the Academy the right—for fifteen years—to publish papers of its members and accounts of its activities "and generally anything the Academy wishes to have appear under its name, *after examining the works and judging that they are worthy* of publication."[8] The highlighted phrase is key to understanding the activities and history of the Comité de Librairie. As the officially sanctioned tribunal for the sciences, the Academy was careful to maintain this privilege.[9]

The Comité de Librairie formed the institutional arm of the Academy that examined and approved works for publication. The Academy's internal rules defined the powers of the Comité. At one point, for example, the Academy voted "to monitor carefully that memoirs are not printed without being read to the Academy and then judged by the Comité."[10] A decision taken in 1721 confirmed that "the Comité has the right to insist on corrections to papers read in the Academy's sessions."[11] About a complicated matter concerning publication that arose in 1786, the pensionnaire botanist, Fougeroux de Bondaroy, recorded the outcome in his own set of minutes and added, "Besides, for members' papers there is always the Comité de Librairie which decides if a memoir read in its assemblies will be printed or not."[12]

The Comité de Librairie constituted one of two standing committees of the Academy, the other being the Comité de Trésorerie, which dates from 1725. The Comité de Librairie functioned essentially without interruption from 1700 until the Academy was disbanded in 1793. In some senses we need to blur the Comité-Académie distinction with regard to censorship powers. For papers read or submitted by outsiders, for example, the full Academy itself usually appointed two or three referees ("commissaires") to review papers, and referees' reports ordinarily included recommendations regarding publication. (These would then go to the Comité with the paper for formal consideration.) Either on appeal from the Comité de Librairie or independently, the Academy occasionally adjudicated a publishing issue sitting as a committee of the whole. Nevertheless, as matters came to be handled ordinarily, the Comité de Librairie evaluated material destined for the *Histoire et Mémoires* and the *Savants Étrangers* series, and it approved and directed papers to the academician-editors of other Academy publications, including the *Connaissance des Temps* and the *Description des Arts et Métiers*.[13]

An institutionalized system thus arose for controlling the publication of French science vested in the hands of the Académie Royale des Sciences and its Comité de Librairie. The Academy's power over the press complemented the rest of the state censorship apparatus, and, indeed, the Academy acted in lieu of other state censorship for much of the science published in eighteenth-century France.[14] Academicians themselves served as ordinary censors and so meshed the two censorship structures. On at least two occasions the Academy voted to demote a work it had previously approved to the world of ordinary censorship so that it would appear without its blessing or acknowledgment.[15] Academicians had the right to publish in that non-Academy sanctioned world, as long as they did not use the title of academician, but the Academy's rules made clear that the institution—and official French science—would have nothing to do with works appearing outside of its orbit.[16]

A thorough comparative study of the Comité de Librairie and publication practices of other contemporary scientific societies exceeds the scope of the present study, but a brief examination of the matter highlights the distinctive attribute of the Paris Academy and the Comité de Librairie to "examine and judge" science.[17] The *Société Royale de Médecine* (Paris, 1776–1793), for example, was a virtual clone of its elder scientific cousin. This society published annual tomes of *Histoire et Mémoires*, and it had a similar Comité de Librairie. As much as it was, in fact, an instrument for control over contemporary French medicine, the Société Royale de Médecine was not prepared or empowered to go as far as the Academy of Sciences in claiming or exercising the right of evaluating the content of medical research and publication. Its *Mémoires* carry the following disclaimer:

> The Society wishes it known that it does not adopt the opinions it presents and that authors of the observations and papers it publishes are the guarantors of the facts they announce. The Company asks that nothing be regarded as institutionally endorsed unless specifically approved by a formal deliberation.[18]

At the *Académie Royale de Chirurgie* (1743–1793), French surgeons articulated an elaborate rationale for a similar "hands-off" policy toward the material they saw into circulation.

> We do not pretend that all the observations we publish are absolutely new. To be sure, one would have to work through all the collections of surgical fact, and even then it is almost impossible that all the circumstances would be the same. Even if they were absolutely identical, we think such reports are worthy of publication because they might be useful or are rare. If useful, the duplicated effort is no problem; if rare, they add to the weight of examples. In addition, . . . in matters concerning the surgical arts, any fact, no matter how old, can at any time be turned into an innovation to improve an art and benefit society.[19]

In the case of the French *Académie de Marine* (Brest, 1752) the king and the superintending Ministry of the Navy and the Colonies were explicit that this learned society devoted to matters naval should *not* possess the powers vested in the science academy. Article 26 of the 1752 Letters Patent for the Marine Academy stated:

> Regarding memoirs submitted by outsiders, the secretary will acknowledge receipt of same without entering into any details. The object of this Company should be only to inform itself and not to raise itself as a tribunal to decide any matter. Nevertheless, the Director . . . can designate academicians to examine the subject, if it appears important, and to draft a reply that they will present at the following meeting. Responses on important matters, approved by the Company, will be sent to the Minister of the Navy who will decide if they should be pursued as a result.[20]

By indirection, the lack of formal publication procedures at the Parisian *Académie Royale des Inscriptions et Belles-Lettres* similarly sheds light on the nature and role of the Comité de Librairie at the Academy of

Sciences. The Academy of Inscriptions (1663) published its own series of *Histoire et Mémoires*, but it had no official publications committee. When time came to choose papers for inclusion in its *Mémoires*, this academy simply appointed an ad-hoc committee. Furthermore, in strong contrast to the Comité de Librairie at the science academy, the review committees of the Academy of Inscriptions included junior academicians.[21]

The Royal Society of London presents the most striking contrast. In 1752 the Royal Society took over publishing the *Philosophical Transactions*, which until then was an unofficial publication of the Royal Society held privately in the hands of its secretary. On assuming control in 1752, the Royal Society created its own Committee on Papers to manage the publication. This committee held regular meetings, kept records similar to those of the Comité de Librairie in Paris, and likewise became the locus of behind-the-scenes, politico-scientific negotiations.[22] While seemingly akin to the Comité de Libraire in Paris, the Royal Society's Committee on Papers felt compelled to insert the following notice in the *Philosophical Transactions* for 1752:

> The Committee appointed by the Royal Society to direct the publication of the *Philosophical Transactions*, take this opportunity to acquaint the public, that . . . it was thought advisable, that a Committee of their Members should be appointed to reconsider the papers read before them, and select out of them such, as they should judge most proper for publication in the future *Transactions* . . . And the grounds of their choice are, and will continue to be, the importance or singularity of the subjects, or the advantageous manner of treating them; *without pretending to answer for the certainty of the facts, or propriety of the reasonings*, contained in the several papers so published, which must still rest on the credit or judgment of their respective authors.[23]

The Royal Society repeated these caveats in print in 1761, and then *annually* in every volume and every series of the *Philosophical Transactions* from 1761 until 1952! Contemporary learned societies—including the Royal Society—controlled their members and the published work of their members in a variety of ways and to varying degrees. That said, the Royal Society's official position regarding its authority over knowledge stands in sharp contrast with that of the science academy in Paris and its Comité de Librairie, which were charged to certify claims and to guarantee truth.

Notes

1. Hahn, *Anatomy*, chapter 1, provides this standard account. For an in-depth examination and case study of changes in research orientation of the Academy in the seventeenth century, see Stroup, *A Company of Scientists*, Part III.

2. Halleux et al., vol. 1, 8–91, documents the production of the Academy in the seventeenth century.

3. Stroup makes this point, *A Company of Scientists*, 59–60, 115–116 and passim. See also her "L'Académie royale des sciences et la censure." I owe a particular debt to Professor Stroup for reminding me that the functions exercised by the Comité de Librairie in the eighteenth century originated in the practices of the Academy in the seventeenth century.

4. Hahn, *Anatomy*, 22–27. Hahn underscores the effective blending of the Academy's role as the judge of technological innovations and contributions to the natural sciences.

5. On the reform of the Academy in 1699, see relevant passages in Hahn, *Anatomy*; Stroup, *A Company of Scientists* and her *Royal Funding*; Sturdy, *Science and Social Status*, Part Four; see also Tits-Dieuaide, "Une institution sans statuts" and her "Les Savants, la société et l'état."

6. See articles XXX and XLVI of the 1699 Letters-Patent in Brian and Demeulenaere-Douyère, eds., *Histoire et mémoire*, 411, 413.

7. The publication privilege was an actual parchment document; the 1717 version is in the archives of the Académie des Sciences, Paris; AdS, PS, 29 juin 1717. See also AdS, DG 31, "Histoire de l'Académie," which holds other documents regarding the publication privilege. The author thanks Professor Stroup for signaling these documents and their importance. Hahn, *Anatomy*, 61, mentions other examples.

8. PS, June 29, 1717: ". . . generalement tout ce que lad. academie voudra faire paroistre sou son nom; apres avoir fait éxaminer lesd. ouvrages et jugés quils sont dignes de l'impression. . . ." Author's emphasis. Quotations in the text from unpublished *manuscript* sources are given in the notes with original spelling and punction. All translations are the author's.

9. Hahn, *Anatomy*, 60–61, makes this point. See also Gillispie, *Science and Polity*, 84, 98; "Privilège du Roy" in first volume of *Recueil des pièces qui ont remporté le prix de l'Académie royale des sciences* (Paris: Claude Jombert, 1732) and "Privilège Général Pour l'impression des Mémoires et autres ouvrages des académiciens, 19 mars 1750" in Aucoc, *L'Institut de France*, xcvi–xcviii. Note the vigorous action of the Comité de Librairie in 1770 to renew the Academy's publication privilege; RCL II/2 (February, 1770).

10. AdS, DG 31 contains several versions in manuscript of the Academy's internal rules. See here the 46-page document dated in a later hand, 1699–1753:

"Collection des Reglemens et Déliberations de L'Académie Royale des Sciences, par ordre de matière," p. 24: "Le 29 juin. L'academie chargea son sécrétaire de veiller exactement à ce qu'aucun mémoire ne fut imprimé sans avoir eté Lû, puis jugé au Comité."

11. AdS, DG 31, "Collection de ses Reglemens et deliberations par ordre de Matieres . . . Redigé par Mr. hellot et Corrigé aprés la lecture qui en a été faite dans les assemblées," p. 28: "Le Comité a droit d'exiger des corrections aux memoires qui ont été lûs dans les Séances." See also AdS, DG 31, "Règlements de l'Académie royale des sciences [1699 to 1738]," fol. 4r, under the date for "1721 Juillet 12," where one reads: "On a délibré sur la manière de faire les corrections que le Comité de la Librairie demande quelques fois à quelques Mémoires qui seront imprimés; et on a reglé que le Secrétaire avertiroit de faire ces corrections, et qu'ils les rapporteroient au Comité suivant."

12. APS Fougeroux papers, "Séance du Samedi. 14 janv.er [1786] . . . il y a toujours pour les memoires des membres le comité de la librairie qui décide si un memoire lu aux assemblées sera imprimé ou ne le sera pas." These personal *plumatifs* taken by Fougeroux at the Academy's meetings from 1786 through 1789 are a rich resource hitherto untapped by historians.

13. See, for example, RCL II/11 (February 1771), where de Milly's piece on porcelain was "renvoyé aux arts," i.e., sent to the Academy's *Description des Arts et Métiers* series; RCL II/18 (January, 1772), where Mallet's stellar tables earned the comment: "à mettre dans la *Connoissance des Temps;*" RCL II/19 (February 1772), where the Comité forwarded Trouson's memoir on mining in Corsica to "M. Duhamel pour estre imprimé à la suite de L'art des forges." Technically, the *Connaissance des Temps* remained independent of the editorial jurisdiction of the Comité de Librairie, and when Lalande once published what was deemed inappropriate material in the *Connaissance des Temps*, the full Academy created a special oversight committee to censor the publication; see PV 84 (1765), fol. 305v.

14. On censorship in pre-Revolutionary France and for an entrée into the literature on same, see Roche, "Censorship and the Publishing Industry." On the censoring and publication of scientific works outside the orbit of the Academy, see case study by Perkins, "Censorship and the Académie des sciences," and Hahn, *Anatomy*, 60. Note that the Academy refused to involve itself when one academician questioned the judgment of another academician acting in his capacity as a government censor; see PV 84 (1765), fol. 141.

15. See APS Fougeroux papers, meetings "du Mercredi 8 mars 1786" and "du Mercredi, 22 mars, 1786," where the Academy voted 21 to 18 to lift its approbation given the work of one de Fer on canals and to consign it to ordinary censorship, "sans que lacademie y soit pour rien." For another decision of this sort, see APS Fougeroux papers, "du mercredi 3 mai 1786."

16. Cf. AdS, DG 31, "Collection des Reglemens et Déliberations de L'Académie Royale des Sciences, . . . 1699–1753," p. 24: "[For works published]

sans l'examen de l'academie . . . il fut décidé à la pluralité des voix qu'il ne devoit estre fait, dans l'histoire, aucune mention d'un tel ouvrage et qu'en cas que l'auteur soit attaqué, il n'a aucun droit de repondre dans les volumes de l'academie." Papers by academicians that were published outside the orbit of the Academy also lost their right to be transcribed onto the registers of the Academy; see PV 99 (1780), fol. 63–63v.

17. The author acknowledges the suggestion of an anonymous referee that a comparative study would clarify the censorship role of the Comité and the Academy.

18. *Histoire et Mémoires de la Société Royale de Médecine*, [Tome 4] 1780 and 1781 (Paris, 1785), 430.

19. Académie Royale de Chirurgie, *Histoire et Mémoires*, vol. 2. [Nouvelle édition avec notes] (Paris: 1819), iii–iv.

20. Cited in Doneaud du Plan, part I, p. 8.

21. See Institut de France, Archives de l'Institut, "Registre des Assemblées & Délibérations de l'Académie Royale des Inscriptions & Belles Lettres," for example, Cote: A: 66, meetings for Friday, 7 December 1770; Tuesday, 28 June 1774; Tuesday, 5 July 1774; Cote A 67, meetings for Tuesday, 30 May 1777; Tuesday, 27 July 1779; Cote A 70, Friday, 7 September 1787, etc. On the domination of the Comité de Librairie by senior academicians, see below, chapter 3.

22. See "Minutes of the Committee of Papers of the Royal Society," Library and Archives of the Royal Society of London, Ms. CMB. 90b.

23. "Advertisement," *Philosophical Transactions* 47 (1751–1752) [London: 1753]. Author's emphasis. For more on the Royal Society's Committee on Papers, see [Royal Society of London.] *Diplomata et statuta regalis societatis Londini pro scientis naturali promovenda. Iussu Praesidis et Concilii edita* (Londoni: Typic Sam. Richardsoni, MDCCLII), 107–110, "Chapter XX: Of the selecting of Papers laid before the Society in order for Publication."

৯ু Chapter Three ৫৩

Statutes and Membership of

the Comité de Librairie

C oincident with the reorganization of the Academy in 1699, a five-person publications committee emerged to assume editorial control over the *Histoire et Mémoires* and other of the Academy's productions. On July 28, 1700 a set of internal regulations recorded: "To speed publication of its memoirs, the Academy resolved that the four members of the Comité de Librairie and the Academy's president will meet on the first of every month to choose which of the works read during the preceding month the Company will publish, with or without changes."[1] How this original committee was chosen or the extent to which it represented continuations from seventeenth-century practices remains unknown. In formalizing the treasurer's right to attend meetings and to vote, a further administrative ruling in the Academy in 1718 raised the number of members on the Comité to six.[2] A later reference called this initial group the "old Comité de Librairie."[3] This committee was responsible for the publication of the Academy's *Histoire et Mémoires* through the volume for 1728 that was published in 1730.

The ad hoc character of the early Comité apparently proved contrary to the highly regulated Academy, for in 1731 Louis XV provided a set of formal statutes for the Comité de Librairie.[4] These rules established a new publications committee consisting of ten "commissaires." Slots were reserved for the Academy's six statutory officers (its president, vice-president, director, sub-director, secretary, and treasurer). The majority of members of the Comité de Librairie thus served ex officio. Two "permanent" seats were set aside for two academician-members elected without term, one chosen from the "mathematical sciences," the other from the "physical sciences," the two main disciplinary classes into which the Academy was divided.[5] Finally, two "annual" at-large positions were filled by other resident academicians who served for a year at a time.

Their service could be renewed, but they could not sit on the Comité for more than two years in a row.

One can only speculate as to the reasons for the expansion—essentially the doubling—of the membership of the Comité de Librairie in 1731. Although at least some of the Academy's officers were members of the old Comité, incorporating the entire group of officers ex officio might have represented moves to tighten up control, to dilute the power of the annual president, to provide greater continuity, or possibly all three motivations. Conceivably, the creation of the four permanent and annual positions was intended to expand representation of the body of academicians. Perhaps the previous committee was perceived as scientifically unbalanced, a move that creating permanent members from the mathematical and physical science classes of the Academy might have rectified. Maybe the workload had increased to the point where more members were needed. More than one factor might have been at play. The record does not clarify these possibilities.

As for how members were chosen, the outgoing president of the Academy annually proposed candidates for election as officers for the following year.[6] Nominations of the at-large annual members and (when the occasion called for it) permanent members probably occurred in a similar way. None of these elections seems ever to have been contested. The Academy kept track of the membership and attended to substitutions, so that a replacement would be found if a permanent committee member was, say, the director or sub-director for a particular year.[7] The regulations of 1731 defined the membership of the Comité and structured its work for the rest of the century.

On the surface, the reforms promulgated in 1776 represented a considerable tightening of procedures affecting the Comité de Librairie. The new rules, issued from the ministry and approved by the Academy, created an additional board of six "Commissaires Censeurs" whose job it was to initial the page proofs of "each volume [of *Histoire et Mémoires*] of the Academy, prefaces to the *Savants Étrangers*, prize volumes, and every other work published in the name of the Academy."[8] At least two of the six censors had to sign each page of proofs, with appeals going to the Comité de Librairie in case they disagreed. Officers were to censor material presented at public meetings, "and authors will be required to make the changes demanded of them. . . ."[9] Disgruntled speakers could appeal to the Comité de Librairie.

Not coincidentally, these drastic new regulations appeared at the very moment the Marquis de Condorcet took over as the Academy's permanent secretary, and they had everything to do with the fears of his enemies and their desire to restrict his field of action.[10] Condorcet was rightly perceived as a liberal aligned with d'Alembert, Turgot, and the *philosophe* cause, and his appointment as assistant secretary, engineered in 1773, deeply divided the Academy and provoked heated opposition from Buffon and other conservatives within the institution in the period leading up to Condorcet's succession as full secretary in 1776. Until this dispute arose the Academy's secretary had enjoyed a free hand in composing material complementary to the scientific papers, such as text for the *Histoire* section of the *Histoire et Mémoires*, eulogies of departed academicians, and prefatory matter for the *Savants Étrangers* series and for the volumes of winning papers from the Academy's prize contests. In his private notes Lalande labeled the new rules "the submission of the secretary to censorship by the officers," and he indicates that the Academy's president and director approved the plan "so that it might seem of general applicability without being directed at the secretary."[11] All in all, the new level of internal censorship that surfaced in 1776 bespeaks the political squabbles within the Academy and little else. Lalande's private notes show that the provisions were soon dropped, in part for want of funds to pay a copyist![12] The statutes of 1731 continued to govern the Comité de Librairie until the closure of the Academy in 1793.

Basing their research on the minutes of the meetings and other records of the Academy, Mmes Pierre Gauja and Christiane Demeulenaére-Doyère compiled a list of who sat on the Comité de Librairie and in what capacity over the period 1731–1793.[13] The work of these past archivists of the Academy makes it easy to sketch a collective portrait of the group.

The most cursory examination of the data reveals the Comité de Librairie to have been dominated by a small group of insiders. The total membership on the Comité numbered only 90 individuals in the 63-year period, 1731–1793, a figure that suggests little turnover on the ten-member panel.[14] The total is also low, considering that the pool of Paris-based academicians who might have been elected (prorated for 1731–1793) is approximately 210.[15] On average, academicians served on the Comité de Librairie for almost seven years, a figure indicating a stable membership and great institutional continuity.[16]

Thirty-four different individuals at least nominally sat on the Comité de Librairie ex officio as presidents and vice-presidents of the Academy.[17] They may be considered to be a separate block of members, at least in part because only *honoraires* of the Academy could be elected president or vice-president.[18] (All but two presidents or vice-presidents in the period 1731–1793 were honoraires. The exceptions were *pensionnaires* Jean Darcet and P. S. Laplace, elected under changed circumstances as president and vice-president for 1793.) Presidents and vice-presidents were overwhelmingly nobles, and with the single exception of Trudaine de Montigny, none ever served on the Comité in any capacity other than as president or vice-president. (Trudaine de Montigny was thrice elected as an annual member.) The custom was for the Academy's vice-president one year to succeed as president the next, and, in fact, the *median* number of years on the Comité for this group is two, a figure suggesting that honoraires served a single sequence as vice-president and president and then ceased any function as an officer in the Academy or on the Comité. The *average* tenure of 3.7 years might imply that presidents and vice-presidents served two rounds in these positions, but this figure obscures an important fact. Although most of the presidents and vice-presidents in the period 1731–1793 did serve only one or two terms as such, a few sat for much longer periods. Louis Phélypeaux, duke de La Vrillière, for example, was an officer for a total of thirteen years; the marquis de Maillebois for ten; the count d'Argenson for eight, and Lamoignon de Malesherbes for eight. (By excluding these four men, whose average tenure approached ten years, the mean number of years on the Comité for presidents and vice-presidents drops to 2.9 years.) La Vrillière and Malesherbes were successively ministers in charge of the *Maison du Roi*, the administrative unit that superintended the Academy of Sciences. Maillebois was lieutenant general of the army; d'Argenson was a Minister of State and Keeper of the Seals. Other repeat vice-presidents and presidents also had strong connections to the government and the crown.

A detailed account of the role of presidents, vice-presidents, and honoraires in the history of the Academy remains to be written.[19] In general we know that Academy presidents and vice-presidents did not participate in the day-to-day life of the Academy, and we may likewise presume that they stayed out of the ordinary business of the Comité de Librairie.[20] Presidents and vice-presidents did intervene in crisis situations, as the case above concerning the 1776 censorship reform reveals. The conflict between Pierre Bouger and C.-M. de La Condamine, examined in Chapter 6, likewise shows the occasional influence of these officers on the Comité de Librairie itself.[21] Most significantly, the present statistics indicate that a handful of honoraires repeated as presidents and vice-presidents for long periods, in part at least to monitor the interests of the crown in the affairs of the Academy and its Comité de Librairie.[22]

Exclusive of presidents and vice-presidents, fifty-eight other academicians served on the Comité de Librairie between 1731 and 1793. This group represents the scientific core of academician-members who ran the Comité's business for six decades. These members filled the various other categories of membership, usually for extended periods, and they typically rotated among different positions. Lavoisier, for example, accumulated a total of thirteen years as director, sub-director, permanent member, annual member, and as treasurer. Only five of this group of academicians were associates (*associés*). The rest were pensionnaires. In other words, a small group of pensionnaires complemented the honoraires in running the Academy and the Comité de Librairie. Composed of individuals active in science, this group was no doubt more influential than the honoraires nominally on the Comité.

Forty directors and sub-directors of the Academy, all pensionnaires, filled these ex-officio slots on the Comité de Librairie in the period 1731–1793. Directors and sub-directors provided the active scientific and administrative leadership for the Academy during the course of a year, and they undoubtedly exercised a similar function on the Comité de Librairie. The *median* number of years for directors and sub-directors on the Comité was two, as it was for presidents and vice-presidents, but unlike the later group, the *average* years on the Comité was just three, and only five individuals served for more than two sets of terms as director and sub-director, none for more than eight years in toto. In other words, although there were stalwarts such as R.-A Ferchault de Réaumur and S.-F. Morand, who filled these positions for eight and seven years respectively, *qua* directors and sub-directors and as a group, these officers did not match the honoraire presidents and vice presidents in terms of longevity on the Comité de Librairie. The point may be moot, in that directors and sub-directors were surely more active and involved than their presidential counterparts.

The permanent secretary of the Academy ran the Comité de Librairie. That much can be concluded from the hand that kept the minutes, as well as from the other, related aspects of the job of Academy secretary. Three of the four permanent secretaries of the Academy in the eighteenth century—Fontenelle, Mairan, Fouchy, and Condorcet—continued on the Comité de Librairie for lengthy periods. Counting from 1731, Fontenelle filled the secretary's position on the Comité for only ten years until his death in 1740, but he was secretary from 1697 and a force behind the Academy's publishing endeavors for more than four decades. Grandjean de Fouchy maintained the register for thirty two years, Condorcet for eighteen. The position did not suit him, so Dortous de Mairan remained secretary for only three years (1741–1743). But de Mairan was nonetheless a fixture on the Comité de Librairie, sitting in various capacities for a total of twenty-eight years. Treasurers likewise served ex officio and were long-term associates. The count de Buffon was nominally a member for twenty-nine years in this role, but his service seems to have been pro forma, in that there is no other evidence of Buffon's impact either as treasurer or, more notably, as the premier naturalist in the Academy. In two instances—for Buffon and for de Fouchy—the Academy appointed assistants, so some ambiguity surrounds the tenure on the Comité de Librairie of secretaries and treasurers, depending on whether the assistant served with or replaced the officer in question on the Comité. Nevertheless, the secretaries and treasurers, and especially the former, constituted the administrative center around which activities of the Comité unfolded and through which the Comité connected with the rest of the Academy.

Permanent members played a distinct role on the Comité de Librairie. They constituted one fifth of the Comité's ten members. They did not serve ex officio, but were elected by their peers. They were all pensionnaires. Strikingly, only eight individuals occupied the two chairs reserved for the permanent members in the whole of the period, 1731–1793. H.-L. Duhamel du Monceau continued as a permanent member of the Comité de Librairie for thirty-nine consecutive years, and with two additional years as a director and sub-director Duhamel ends up with the overall record of forty-one years on the Comité. Réaumur was another long-standing permanent member, serving for twenty-seven years. The other six permanent members of the Comité de Librairie were Jean le Rond d'Alembert (12 years), Jean-Charles de Borda (7 years), Pierre Bouguer (1), Antoine-Laurent de Lavoisier (8), Jean-Baptiste Dortous de Mairan (13 years), and Louis Lémery (13 years). The average tenure of a permanent member was fifteen years, a figure that rises to seventeen years if one discounts Pierre Bouguer who died the year he became a permanent member. Plainly, this core of permanent members, along with the secretaries, added great continuity to the Comité's business and consciousness.

The number of annual members elected to the Comité de Librairie in principle could have totaled 126, but in fact only thirty-four individuals filled these two seats over the sixty-three year period in question. In other words, annual members were chosen from a relatively small pool of academicians, and they were elected more than once, but not much more than once. The median stint was three terms, and the average tenure for annual members rises slightly to 3.3 years. That figure argues that annual members were not long-term members like their permanent counterparts or the Academy's officers, but instead sat on the Comité for comparatively short periods of time. Twenty-eight of the thirty-four annual members were already pensionnaires; the six others were the previously mentioned associate members, four of whom later became pensionnaires. There was considerable overlap among annual members, permanent members, and academicians who rotated through as directors or sub-directors. Nine known pensionnaires replaced sitting members of the Comité de Librairie when one of the latter became directors or sub-directors for a term. The profile of directors, sub-directors, and annual and permanent members reinforces the picture of the Comité as governed by a small coterie of the Academy's top-ranking members and officers. Incidentally, it seems that many academicians filled a term on the Comité as an annual member just after election as pensionnaire, that is, a year on the Comité was the norm for a new pensionnaire.

The same portrait of the Comité de Librairie as a restricted group of Academy insiders emerges when one looks at the members of the Comité, not from the point of view of the seat they occupied on the Comité, but of the rank they held in the Academy itself.[23] Pensionnaires made up over half of the membership of the Comité, honoraires over a third. When one adds the secretaries and treasurers, the number of honoraire, pensionnaire, and officer members of the Comité approaches 95%. Put another way, only a tiny fraction of the associate academicians and none of the most junior members of the Academy—its *adjoint* or adjunct members—was ever represented on the Comité. Similarly, no one from the Academy's non-resident categories of free associate, foreign associate, or correspondent ever had a direct voice on the Comité. The correlations between rank in the Academy and position on the Comité have already been pointed out: honoraires were *de jure* presidents and vice-presidents, and pensionnaires held every other position on the Comité with the exception of the few associates who on occasion were elected as annual members. The preponderant role of pensionnaires is further indicated by their average length of service on the Comité, a comparatively long 7.5 years.

Finally in this connection, pensionnaires and associates who served on the Comité de Librairie represented different areas of scientific expertise, and, as it turns out, they nicely balanced the disciplinary divisions into which the Academy as a whole was organized.[24] The Letters Patent of 1699 structured the Academy into three sections for the "mathematical" sciences (geometry, mechanics, and astronomy) and three sections for the "physical" sciences (anatomy chemistry, and botany).[25] The organizational reform of the Academy in 1785 added experimental physics to the mathematical sciences grouping and natural history to the physical sciences grouping. In a nontrivial number of cases the scientific expertise of academicians did not completely correspond to the disciplinary class to which they had been elected. For example, d'Alembert started out as an adjunct in astronomy before becoming an associate in geometry, a pensionnaire in mechanics, and finally a pensionnaire in geometry. Daubenton moved up the ranks from adjunct and associate in botany to associate and pensionnaire in anatomy. J.-B. Le Roy passed from adjunct in geometry to associate and then pensionnaire in mechanics, and then pensionnaire in experimental physics with the reform of 1785.[26] The formal disciplinary divisions of the Academy thus tracked, but did not exactly mirror the scientific fields of individual academicians, and hence members of the Comité de Librairie may not have truly represented the disciplinary class with which they were associated in the Academy. That subtlety notwithstanding, it is notable that the mathematical and physical science groupings of the Academy are exactly balanced among the pensionnaire and associate members of the Comité de Librairie. The distribution of members in the disciplinary subdivisions of the mathematical and physical sciences was somewhat uneven, with astronomy and anatomy having comparatively few representatives on the Comité de Librairie. No one from experimental physics can be found on the Comité from the inauguration of the class in 1785 through 1793. Nevertheless, the membership of the Comité de Librairie in general reflected the scientific orientation of the Academy as a whole, and it seems reasonable to suggest that these considerations played a role when members were chosen for the Comité de Librairie.

Notes

1. AdS, DG 31, "Règlements de l'Académie royale des sciences . . . 1699 à 1738," fol. 1r: "Pour hâter les affaires qui regardent la Librairie, et l'impression des Mémoires de l'Académie, il a été résolu par deliberation prise à l'ordinaire, que chaque premier jour du mois les quatre Commissaires de la Librairie présidés par M. le Présid.t s'assembleroient et régleroient quels ouvrages entre ceux qui aurient été lus les mois précéd.t seroient imprimer par ordre de la Compagnie, avec des changemens ou sans changemens."

2. Lalande, "Collection," 46, under rubric "Tresorier."

3. "Ancien comité de la Librairie." AdS, DG 31, "Collection de ses Reglemens et deliberations par ordre de Matieres . . . Redigé par Mr. hellot et Corrigé aprés la lecture qui en a été faite dans les assemblées," marginal notation, p. 47.

4. See PV, vol. 50 (1731) 177, which transcribes the ministerial letter dated "A Fontainebleau le 17 Juillet 1731." See also PV 50 (1731), 181.

5. See Hahn, *Anatomy*, 98–101; Gillispie, *Science and Polity*, 91, and below, Appendices.

6. See Lalande, "Collection," unnumbered page, where Lalande notes: "En 1780 l'usage est que le president actuel propose au ministre les officiers pour l'année suivante." See also Keith Baker letter of September 9, 1966 in AdS DG 31, which discusses these matters. The author thanks Professor Roger Hahn for raising the question of how members of the Comité de Librairie were chosen.

7. See APS Fougeroux papers, "Samedi 4 aoust 1787," "du Samedi 12 jan.er [1788]," "Mercredi le 7 jan.er 1789," where Fougeroux notes elections and substitutions to the Comité de Librairie.

8. See, "Projet reglant impression des M," PS, January 10, 1776: ". . . six Commissaires Cenceurs, les quels parapheront avant le tirage, les feuilles imprimées de Chaque Volume de l'academie, celles des préfaces des Volumes des Sçavans etrangers, des Volumes des prix et Celles de quelque ouvrage que ce soit, publié sous son Nom."

9. "Projet reglant impression des M," PS, January 10, 1776: ". . . et l'auteur sera obligé par provision de se Conformer aux changemens qu'ils pourront exiger. . . ."

10. See Baker, "Les Débuts de Condorcet," and his *Condorcet*, 35–47. See also, Lalande, "Collection," 125.

11. Lalande, "Collection," 8: "Secretaire soumis a la censure des officiers pour les éloges, et pour l'histoire de l'academie a celle de 6 commissaries censeurs, suivant un reglement de 1776 rédigé par m. d'alembert, corrigé par divers comités tenus chez m. Tenon directeur et m. de maillebois president pour que le reglement paroisse etre general sans toucher spécialement au Secretaire."

12. Lalande, "Collection," 8: "Les commissaries censeurs verifieront le registre tous les 3 mois. Depuis plusieurs années il est interrompu a defaut de payemens du copiste."

13. AdS, DG 31, "Comité de Librairie d'après le Règlement de 1731." See discussion below, Appendices. The minutes of meetings of the Comité de Librairie shed no light on the membership; see RCL, passim. The author thanks Mmes Marie-Jeanne Tits-Dieuaide and Claudine Pouret for their help in further identifying the membership of the Comité.

14. See Appendix 1(a).

15. Extracted from McClellan, "A Statistical Portrait," here 546–47.

16. See Appendix 1(b).

17. See Appendix 1(c).

18. Gillispie, *Science and Polity*, 82.

19. For a social analysis of the scientific members of the Academy through 1750, see Sturdy, *Science and Social Status*.

20. Hahn, *Anatomy*, 77.

21. See below, chapter 6.

22. Hahn, *Anatomy*, 80, remarks that ministers of state superintending the Academy were "automatically elected" as honoraires when positions became open.

23. See Appendix 1(d).

24. See Appendix 1(e).

25. See above at note 5 and below, Appendix 1(e) and its explanatory note.

26. See the *Index biographique* of the Académie des Sciences, passim. In 1739 Buffon famously passed from the mechanics section to the botany section, but this move represented a genuine shift of scientific and career interests on Buffon's part.

᧖ Chapter Four ᧗

What Got Published, What Did Not,

and Why

The Comité de Librairie functioned as a gatekeeper, carefully winnowing what would and would not appear in the Academy's publications. The Comité had the day-to-day responsibility of sifting through an
endless stream of papers, letters, memoirs, accounts, notes, observations,
remarks, complaints, and so forth by academicians and others passed on
to it by the full Academy through the office of its secretary. The Comité
had to keep track of these submissions and evaluate their suitability for
publication. That evaluation required procedures, the most important
detailed in this chapter being the institution of peer review. In deciding to
approve or reject any particular piece, the Comité had to exercise judgment and apply standards. Those standards seem self-evident and elementary to us today, but it was their consistent application over the course
of a century that helped transform them into common scientific practice.
Then, as now, papers had to conform to a host of unstated parameters that
made them recognizable and acceptable as contributions to the science of
the day. More in particular, the Comité required that papers concern
science and not have been published elsewhere, that they embody new
knowledge and represent original work, that they cite previous research,
that they be civil, and so on. The members of the Comité held substantive
scientific commitments that came into play in assessing works, and as
subsequent chapters will show, the Comité was not immune from partisan
influences. Yet, without reifying the body, the Comité de Librairie was
nevertheless an established forum that transcended somewhat the particular views of its members or the reigning concepts of contemporary
science. As the Academy's agent and instrument, the Comité de Librairie
exercised the institutional privilege to "examine and judge" by sanctioning
works that appeared under the imprint of the institution. Except in rare
occasions when it was overruled or otherwise directed by the Academy as

a whole, the Comité de Librairie was in many ways the effective and ulti-
mate arbiter of the science produced by the Academy. The practical
importance of the Comité de Librairie for French and European science
in the eighteenth century cannot be underestimated.

The Comité de Librairie generally met once a month while the
Academy was in session, and the Academy's secretary kept a separate set
of its minutes. These have been preserved in their entirety for the years
1749 to 1780 in the two folio manuscript registers previously mentioned.[1] These registers total 268 manuscript pages, and in them one can
observe in detail the customary operations of the Comité over three
decades. A typical entry is sparse: it indicates the date of the meeting (but
not attendance) and the month being reviewed by the Comité. Pages are
divided into two columns. (See Figure 5.) The right-hand column lists the

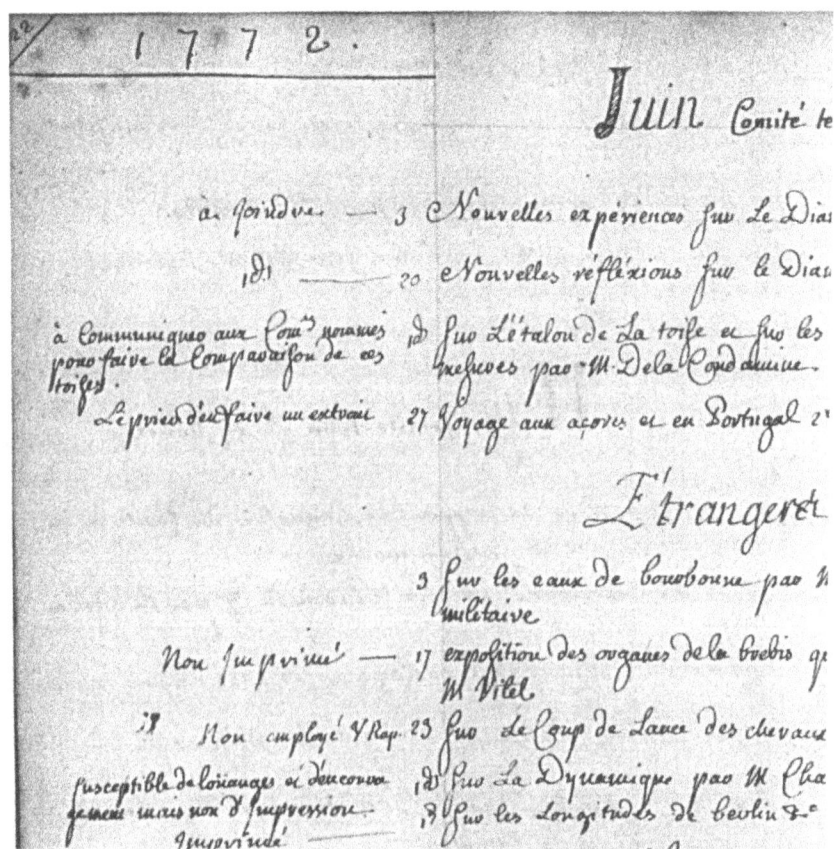

Figure 5: Minutes of the meeting of the Comité de Librairie of July 24, 1772
held to evaluate work presented in June 1772. RCL II/22.

papers and other items being considered, arranged under the rubrics of "Mémoires" (for the papers of academicians), "Savants Étrangers" (for formal papers submitted by outsiders), and "Observations" (for miscellaneous submissions). The left-hand column records the Comité's decisions (publish, send back to author, etc.). Occasionally, a more substantial entry addresses some other, usually controversial matter. Taken together, the unexceptional entries and the extraordinary ones provide remarkable insight into the realities of conducting institutional business and the forces and factors at play in determining the science the Academy made public under its aegis.

Memoirs and reports had to be read to the full Academy prior to consideration by the Comité de Librairie. Items before the Comité then had to meet certain unstated but definite criteria even before they underwent formal scrutiny. In the first instance, papers had to be the work of the nominal authors. The exception made for J.-D. Maraldi's 1788 observations of the satellites of Jupiter, actually performed and written up by his nephew, underscores this elementary principle of authorship and assigning credit.[2] Then, papers had to concern science. The Comité returned Charles-Marie de La Condamine's account of his trip to Italy, for example, with the request to "choose that part which pertains to the Academy's business."[3] Papers also had to be complete and actually submitted, as is evident, for example, when the Comité put off evaluating Joseph-Nicolas Delisle's paper on solar eclipses until it was submitted, or when the Comité similarly rejected Michel Adanson's piece on a Senegalese torpedo as "incomplete."[4] Works could not be too long, as Jean-André de Lucques learned when the Comité turned down his piece on the atmosphere because "it is a book and not a memoir."[5] Anonymous submissions were routinely rebuffed, as were simple announcements, prefaces, or prolegomena to full works.[6] Papers also had to embody new knowledge. Gallon's memoir on an unusual pump in Brussels, for example, was dismissed with the terse, "Nothing new, do not print."[7] Papers likewise could not be too elementary, as G.-J. Le Gentil found out when asked to eliminate parts deemed jejune from his memoir on the apparent diameter of the sun.[8]

More than any other factor, the Comité insisted that papers be original and not published elsewhere. For example, it turned down Grichow's observations from the island of Oesel in the Baltic because "the author published this at Petersburg."[9] Adanson's paper on the baobab tree was excluded because it was "taken from his book."[10] The Comité refused Vicq d'Azyr's first paper on the comparative anatomy of fishes, saying it was "the same as the one published in the volume for 1772."[11] The Comité even cautioned the secretary not to reprint his own eulogies of academicians before these appeared the *Histoire et Mémoires*.[12]

The Comité sought to ensure its exclusive access to papers. Regarding Delisle's memoir on lunar eclipses, for example, it elected "to wait and see if he has it published elsewhere." On the same author's method to determine the return of comets, the group decided to "wait and verify that it is not [already] published."[13] In a revealing instance in 1764 E.-J. Bertin submitted his second and third memoirs on fetal blood circulation, the first of which had appeared in 1753. Suspicions arose that the author was plagiarizing himself. The Comité appointed J.-R. Tenon to compare the 1753 piece with the later submissions. Tenon found nothing untoward in them, but the Comité asked him to review the page proofs of Bertin's papers just in case.[14]

The Comité de Librairie did not receive any papers that concerned religion or politics, but other subjects were no less forbidden. The Comité refused to countenance secrecy, for example. It postponed consideration of academician F.-D. Hérissant's powder for drying and preserving animals "until he gives its composition," and, conversely, it seemed eager to publicize the secret of Delachevalleraye's powder.[15] The Comité likewise evidenced great skepticism regarding purported "causes" or "proofs." It allowed L'Outhier's report on an earthquake to be noted in the *Histoire* section, for example, but "without mention of the proof." Regarding E.-F. Dutour's "Consideration of the Causes of Refraction," the Comité asked the author to "reduce this memoir to the facts."[16] The Comité had little patience for reports of monstrous births, and it rejected the one submission that came to it regarding the squaring of the circle.[17]

Different fates awaited papers meeting the threshold criteria. Outright rejection ranged from the scornful "nothing" ("néant") to the gentle let down that awaited J.-A.-C. Charles' paper on dynamics: "Worthy of praise and encouragement but not publication."[18] Resident academicians received somewhat softer treatment than outsiders, as was the case for the new supernumerary adjunct in chemistry, one Antoine Lavoisier, who was politely asked to "re-read" his memoir on the density of fluids.[19] In some opaque instances the Comité "spoke" to authors privately about their papers.[20]

Before accepting papers for publication the Comité typically had them refereed, and peer review constituted a powerful means by which the Comité influenced the production of knowledge. Regarding an astronomy paper by Cassini de Thury, for example, the Comité asked Clairaut and Bailly "to read it and give the Comité their opinion regarding it." A follow-up notation remarked, "Mr. Bailly's report subsequently seen, publish."[21] Le Roi and Lavoisier were appointed referees for Masnier's experiments on electricity, which the Comité ultimately chose to "publish with the qualifications of the report."[22] Lafosse's paper on stanching hemorrhages was to have been published along *with* the referee's report, but the author withdrew the paper instead.[23] Father Chevalier's observations on lunar eclipses of 1755 and 1757 were "printed with a note tailored to the report."[24] Sage's paper on gold found in ashes was "sent back, given the judgment of the commissioners."[25] These examples evidence potent institutionalized mechanisms for reviewing the scientific content of papers. Crucially, the referees were not lay authorities, or even experts operating in the name of lay authority (as was the case of ordinary censorship in contemporary France), but scientific peers in an institution devoted to science and controlled by the producers of knowledge themselves. If this instance was not the absolute origin of peer review in science, it was certainly an early and weighty episode.[26]

As a feature of the review process, the Comité de Librairie did not simply reject or approve papers, but took steps to verify claims, demand evidence, and ask for revisions. Bordenave's analysis of bile, for example, elicited the comment: "Wait for the experiments demanded by the report."[27] Regarding experiments by Fathers Bertier and Cotte, the Comité opted "to wait until these have been repeated with greater care."[28] Lavoisier's analysis of gypsum received approval, but only if "communicated to the referees before being printed."[29] Regarding Bézout's paper on the three-body problem, the Comité ruled: "Print on the condition that [he] immediately submit corrections."[30] These examples could be multiplied.

The norm of citing prior research had not become commonplace in the world of eighteenth-century science, and the Comité regularly demanded revisions to acknowledge the work of others. Navier's paper on dissolving mercury in acids, for example, was approved with the proviso that he "add the citations mentioned in the referee's report." About a paper on the magnetic needle, Lalande was "asked for clarifications and citations." The Comité similarly requested Nollet "to cite what has been said on this topic," that of small fish becoming agitated before a storm.[31] Regarding Courtivron's method for observing the height of the north celestial pole, the Comité made a point of asking the author "to cite work on the same subject by Mr. de Mairan read and published in 1736," an entry suggesting that the Comité was attentive to the Academy's own place in history.[32] These cases and many others that could be cited show the Comité de Librairie to have been an instrument by which the norm of citing prior work became established.[33]

Authors responded to the Comité and made changes in their papers. One wrote directly to a referee: "I received the report containing the clarifications you ask for to make my work less imperfect. I have nothing more at heart than to satisfy you. I have made . . . the additions and changes you suggested to make the paper more interesting and useful, and I have the honor of sending a revised version."[34] In one instance an aggrieved author, one Pereire, appealed a negative decision by the Comité through the offices of the Academy's director. The latter resubmitted the memoir and supporting documents to the Comité, which decided that, "Before anything else, these are to be communicated to Mr. Bézout who drafted the [original] report in order to take suitable steps after his response to settle this matter." At a special meeting seven months later, the Comité "resolved that Mr. Pereire's memoir could not be published in the fifth volume of the *Savants Étrangers* except in return for corrections noted by Mr. Morand *folio and verso* [sic], that Mr. Morand be requested to review the printer's proofs, and that if Mr. Pereire does not agree, his memoir will not be published."[35] Pereire complained again when his paper was not scheduled for inclusion in the fifth volume of the *Savants Étrangers*. The Comité finally agreed to include a revised version in that volume, "After it is reread by the original commissioners, however."[36] Pereire's paper on the deaf and dumb does appear in the fifth volume of the *Savants Étrangers* series, where he alludes to its abridgment and to the Academy's intervention.[37]

Given the scrutiny to which papers were subjected, authors sometimes took back papers rather than face rejection.[38] Custom had it that works under consideration were withdrawn on the death of an author, and the Comité returned papers to newly elected academicians who had submitted them as outsiders, so that they could be re-read to the Academy and published in the *Mémoires*.[39] In a few odd instances papers were "approved but not published."[40]

The *Histoire* section of the *Histoire et Mémoires* was the repository for miscellaneous scientific reports and observations, and the Comité selected what appeared there. The notice concerning a needle that someone swallowed and rendered anally seems typical and received the fitting comment: "Histoire, very short."[41] Regarding Lieutaud's curiosities from the Île de France sent to J.-E. Guettard, the Comité "asked Mr. Guettard to rewrite an abstract of these for the history section." It similarly "asked the abbé Nollet to provide a notice of Waitz's letter on electricity for the *Histoire*," and it decided to "mention [Foucroy's observations of barometric variations from Calais] in the *Histoire* section . . . and to ask Mr. Lavoisier to reduce them to tables."[42] These examples demonstrate that the Comité and other academicians besides the Academy's secretary had a hand in crafting the annual history of the Academy. The registers are explicit on this point when they refer to "Messrs. Delandes, Tillet, Le Roy and Baron [being] charged with writing the history section for the years 1757, 1758, 1759 and 1760."[43] That many hands were involved sheds a new light on the *Histoire* section and the Comité's role in crafting it.

Two other categories of Comité action deserve notice at this point. One, authors sometimes sought approval for publication outside the orbit of the Academy but with the Academy's endorsement. For example, the Comité granted J.-N. Buache de la Neuville its "approbation" for the separate publication of his map on the southern ocean, as it did Charles Baër for his translations of memoirs of the Swedish *Kungl. Vetenskapsakademie* on epizootics.[44] Simply to make presents of a limited, private edition of his memoir on mapping the Caspian Sea, the royal geographer and Academy associate, J.-B. d'Anville, first had to get permission from the Comité de Librairie and from the Academy's printer, who might have objected to the competition.[45] The imprimatur of the Academy doubtless reflected well on the credibility of the works in question and helped improve sales.

Secondly and finally in this connection, the Comité de Librairie also supervised the publication of papers from the *Société Royale des Sciences* in Montpellier that appeared in the *Histoire et Mémoires* of the Paris Academy. From its inception in 1706 the Montpellier society enjoyed the statutory right of publishing a paper annually in the *Mémoires* of the Paris Academy. The academicians in Paris did not automatically accept the work of their lesser southern cousins, however. Papers from Montpellier had to be read before the full Academy in Paris and were subject to being refereed just like any other work. The paper by the Montpellier doctor Charles Le Roy on atmospheric water vapor, for example, received permission to appear in the Paris volume for 1751, but "with the corrections that the Academy has indicated."[46] In 1765 and again in 1771 the Comité in Paris rejected submissions from Montpellier and asked for "something more suitable" from E.-H. de Ratte, the secretary of the Montpellier Society.[47]

All in all, the record shows that the Comité de Librairie vigilantly monitored and reviewed the materials that came before it. It defined and maintained the set of standards outlined here for what constituted work acceptable for publication. On behalf of the Academy, the Comité de Librairie conscientiously performed its functions to "examine and judge" the work of academicians and outsiders and so to guarantee the truth of the science the Academy published. Subsequent chapters will show what else the Comité de Librairie did to shape the content of what appeared in the Academy's pages.

Notes

1. RCL I and II; see description above, Abbreviations. For unexplained reasons the Comité met only once, on March 20, 1771, to consider all papers submitted from March 1770 through March 1771; i.e., the Comité did not meet for a year; see RCL II/2 (March 1770) through RCL II/11 (February 1771).

2. The elder Maraldi was on his deathbed at the time. See the paper, communicated by J.-D. Cassini IV, MARS (1788), 718–21 and comments MARS (1788), xv, 718, and PV 107 (1788), fol. 187 where the Academy directed that the younger Maraldi's observations be "printed in the *Mémoires* of the Academy as [if] done under the guidance and the direction of his uncle." ["Imprimé dans les Mémoires comme faites sous les yeux et sous la direction de son oncle."] The nephew, Jacques-Philippe Maraldi, was not a member of the Academy and technically not entitled to publish in the *Mémoires*.

3. RCL I/63 (April 1757): "Le prier de faire un choix de ce qui appartient à L'objet de L'académie." La Condamine's paper, "Extrait d'un Journal de voyage en Italie," appears in *MARS* (1757), 336–410.

4. See RCL I/4 (July/August 1749); RCL I/47 (August 1755), "pas achevé."

5. RCL I/114 (May 1762): "C'est un livre et non un mémoire."

6. For rejected anonymous works, see, for example, RCL I/78 (November 1758) and I/147 (August 1765). On project proposals by d'Alembert, Buache, the abbé Chappe, the Marquis de Courtanvaux, and Father Pingré, see RCL I/149 (December 1765), I/178 (September 1768), I/167 (November 1767), I/168 (December 1767); I/88 (November 1759), all of which were rejected.

7. RCL I/2 (March 1749): "Il n'y a rien de nouveau; point d'Impression." See also RCL I/65 (July 1757), where the Comité similarly turned down a report on spontaneous combustion because "M. Duhamel en a donné un mémoire."

8. RCL I/48 (November 1755): "Imprimer et le prier de retrancher des choses trop élémentaires pour L'Impression."

9. RCL I/35 (November 1753): "Lauteur L'imprime à Petersbourg."

10. RCL I/76 (July 1758): "est tiré de son livre."

11. RCL II/34 (September 1773): "est Le même que celuy du 24 Novembre imprimé en 1772 v."

12. RCL I/104 (April 1761).

13. For these examples, see RCL I/58 (August 1758): "Attendre et scavoir s'il ne le fait pas imprimer ailleurs." RCL I/83 (May 1765): "Verifier s'il n'est pas imprimé."

14. RCL I/132 (January 1764); I/148 (November 1765).

15. For the Hérissant episode, see RCL I/96 (May 1760): "Attendre qu'il en ait donné la Composition." For Delachevalleraye, see I/17 (June 1751) and HARS (1751), 82–83. See also the case of Isnard's ether; RCL I/25 (November 1772) and PV 92 (1773), fols. 45–47, where "l'académie en louant le zèle de l'auteur peut l'inviter à communiquer son procédé."

16. On L'Outhier, see RCL I/12 (December 1750): "Histoire sans parler de la preuve." On Dutour, see RCL I/174 (April 1768): "Le prier de reduire ce mémoire aux faits."

17. On monsters, see RCL I/13 (January 1751), I/14 (March 1751), I/46 (July 1755), and Tort, *L'ordre et les monstres*. On the quadrature piece, see RCL I/162 (February 1767). The Academy formally refused to accept any more papers dealing with quadrature in 1775; see Hahn, *Anatomy*, 143–147.

18. RCL II/22 (June 1772): "susceptible de loüages et d'encouragement mais non d'Impression."

19. RCL I/182 (February 1769); this paper was not published in Lavoisier's lifetime.

20. See, for example, RCL II/[72] (August 1778), where Le Gentil's letter on phosphorescence elicited the comment: "En parler à L'auteur."

21. RCL I/133 (February 1764): "Prier MM. clairaut et bailly de le lire et d'en dire leur avis au Comité. Depuis vû le rapport de M. bailly imprimé."

22. RCL II/12 (April 1771): "Imprimé avec les reserves du Rapport;" see also PV, 90 (1771), fol. 91; this paper seems never to have been published.

23. RCL I/12 (November 1750).

24. Re Chevalier, see RCL I/65 (July 1757): "Imprimé avec une note conforme au rapport. [Then in a later hand:] 4e Vol.e." Chevalier's paper, "Observatio Eclipsis Lunæ" appears in Latin in SE, vol. 4: 281–284, accompanied by a note in French clarifying differences between Chevalier's observations and those of one M. de Barros in Paris, having to do with observing the moon through different colored glass plates, weather conditions at the time of the observations, and the different telescopes involved.

25. See RCL II/[72] (August 1778): "renvoié après le jugement des commissaires." Sage's paper became the subject of a separate review ordered by the Academy and conducted by the chemistry section, the reason being that the possibility of finding gold in ashes might tempt the foolhardy; see HARS (1778), 25–27; MARS (1778), 548–559. Lalande's notes indicate a further sequela in 1780: "Denontiation d'un

livre de M. Sage ou il y avoit des choses contre M. Tillet et contre le jugement de l'a-cademie sur l'or des vegetaux, 13 dec. 1780. on a pretendu que ce n'etoit point la une fonction du secretariat;" Lalande, "Collection," 125 under rubric "Denonciations."

26. To the author's knowledge, the history of peer review in science has not been systematically investigated, so the extent to which peer review in the Academy and the Comité de Librairie was a completely novel development remains a some-what open question. That uncertainty notwithstanding, the practice of peer review in the eighteenth-century Academy clearly represents a key stage in the history of peer review. Moran, *Silencing Scientists*, sheds some light on peer review in historical con-text. While concerned with peer review in contemporary science, the books by Speck, *Publication Peer Review*, and Daniel, *Guardians of Science*, disclose contentious issues at play in the peer-review process today, many of which, such as the status of authors, would seem operative in the eighteenth century.

27. RCL I/154 (May 1766): "Attendre les expériences demandés par le rapport."

28. RCL II/41 (June 1774): "attendre qu'elles ayent été plus exactement répétées."

29. See RCL I/142 (February 1765): "Imprimé mais Communiqué aux Comm.es avant L'impression."

30. RCL I/84 (June 1759): "Imprimé a Condition de remettre incessamment les Corrections."

31. Re Navier, see RCL I/93 (May 1760): "Imprimé et ajouter les citations Conformmément au Rapport." Navier's paper appeared in SE, vol. 6 (1774), 325–50. Re Lalande, see PV 91 (1772), 244 and RCL II/23 (July 1772): "hist. et demander a M. de la Lande les eclaircissements et les citations." Re Nollet, see RCL I/146 (June 1765): "histoire et le prier de citer ce qui a été dit sur cette matière." About the Nollet piece the registers add the odd note: "Depuis n'en pas par-ler à la priere de L'auteur."

32. RCL I/44 (March 1755): "faire mention du même de M. de Mairan sur la même matière lû en 1736 où il est imprimé." Newly appointed academicians were expected to read the registers of the Academy (its *procès-verbaux*) and refer to these when appropriate; Professor Stroup, private communication.

33. See Mulkay, "Norms and Ideology in Science," and discussion in Chapter 1.

34. HBI, ms. AD 215; ALS, Jean-François de Marcorelle to Michel Adanson, dated "A Narbonne le 9 Janvier 1766": "Mr de fouchi . . . me fit passer de Votre part et de Celle de M. de Jussieu Un ecrit Contenant des Eclaircissements que Vous Croiés necessaires Et que Vous demandiés pour rendre mon travail moins Imparfait. Je n'ai rien tant à Coeur que de remplir Vos Vues. Jai fait . . . Les additions et Les changements que vous avés estimé Pouvoir le rendre plus Interessant et plus Utile. J'ai l'honneur de L'envoïer sous La Nouvelle forme. . . ." See also companion letter, HBI, ms. AD 216.

35. See RCL I/121 (January 1763): "M.r Le Directeur . . . a remis en même temps Les mémoires et les pièces presentés par M. Pereyre il a été décidé qu'avant tout ces memoires seroient Communiquée à M.r bezout qui a redigé ce rapport pour prendre après sa réponse les mesures Convenables à L'Arrangement de cette affaire." RCL I/156 (August 1766): "Il a été decidé que le mémoire de M. Pereyre ne seroit imprimé dans Le 5e Volume des scavants étrangers que moyennant Les Corrections y faites par M. Morand qui La paraphé au *folio et verso* que M. Morand seroit prié de revoir les épreuves Luy même tous de L'Impression et que si cela n'étoit point accepté par Mr. Pereyre son mémoire ne seroit point Imprimeé. . . ."

36. RCL I/153 (April 1766). "après cependant qu'il auroit été relu par Mrs. Les Comm.res qui ont pris Connoissance. . . ."

37. SE, vol. 5 (1768), pp. 500–530 and allusions, p. 500.

38. See, for example, RCL I/3 (April 1749), I/8 (March 1750), etc.

39. On papers withdrawn on an author's death, see RCL I/8 (March 1750); I/17 (July 1751). On papers returned to authors after their election as academicians, see RCL I/21 (January 1752); I/28 (November 1752); II/71 (March 1778); PV 89 (1770), fol 307v. Only bona fide academicians had the right to publish in the *Mémoires*.

40. See, for example, RCL I/128 (August 1763): "Non Imprimé quoy qu'approuvé."

41. RCL I/30 (March 1753): "histoire. Très court."

42. RCL I/13 (February 1751): "Prier Mr. Guettard d'en de recrerier un extrait pour Lhistoire ou pour le Volume des étrangers." RCL I/75 (June 1758): "Prier M. L. Nollet d'en donner un extrait pour Lhistoire." RCL I/180 (December 1768): "en parler dans L'hist. de même que de celles de Cadix et prier M. Lavoisier de les réduire en tables." See also HBI, AD 266, where the Academy's secretary, de Fouchy, requested a notice for the *Histoire* section prior to the publication of the formal paper in the *Mémoires*, and indeed mention of discovery of a new species of barley appears in the *Histoire* for 1764, a year before Adanson's paper is published in the *Mémoires* for 1765; compare HARS (1764), p. 77 and MARS (1765), pp. 613–619.

43. RCL I/129–30 (June 1762): "Mrs. Delandes, Tillet, Le Roy et Baron Chargés de la redaction de L'histoire des années 1757, 1758, 1759 et 1760." In another instance P.-C. Le Monnier complained about an abstract written by Lalande for the history section, "sur quoi il a été décidé que L'extrait ne paroîtroit point et que le mémoire seroit seulement indique dans L'histoire comme renvoyé entièrement aux mémoires;" RCL I/115 (June, 1762).

44. The standard phrase was "Permis d'imprimer à part sous le privilège de l'Académie." On the cases mentioned here, see RCL II/48 (February 1775) and II/57 (February 1776). Note that the Academy's imprimatur was given only to works to be published in France; it did not apply to editions printed or pirated abroad; on this point see PV 99 (1780), fols. 39–44v.

45. D'Anville was successful; RCL II/69 (May 1777). In 1777 the Imprimerie Royale was the printer of the Academy's *Histoire et Mémoires*.

46. RCL I/38 (March 1754): ". . . seroit imprimé dans le Volume de 1751 avec les corrections que Lacade. a indiquées." This, despite the fact that the author's older brother, J.-B. Le Roy, was an academician in Paris. The paper appears in MARS (1751), 481–518; see also PV 73 (1754), 63.

47. RCL I/144 (April 1765): "J'ay été chargé D'ecrire à M. De ratte pour le prier d'engager . . . Mrs. Dela Société royale à en envoyer un autre pour 1763." RCL II/15 (August 1771): "J'ai été chargé d'en demander un autre plus Convenable." For other instances dealing with Montpellier, see RCL I/4 (July/August 1749); I/14 (February 1751); I/29 (January 1753); I/99 (November 1760); I/186 (June 1769); II/46 (December 1774); and PV 92 (1773), 3.

ᏸ Chapter Five ᏺ

Establishing Priority

*T*he publication of a scientific paper is an invitation for others to freely take and use the information it contains, and the producers of knowledge of nature customarily give away their work in exchange for the credit they accrue.[1] This credit is nonpecuniary, but it represents the essential currency that gauges accomplishment and builds reputations in the sciences. Securing proper credit for scientific work, therefore, is of the utmost concern to the parties involved, and credit is assigned to the individual who first makes a discovery or achieves a result. Establishing priority is an essential and potentially contentious element of the process of making knowledge in the sciences, and those involved have seemingly always kept an anxious eye out to guard their interests.

Chapter 1 noted some of the ways priority was commonly established before the advent of scientific societies in the seventeenth century: publication in books, results circulated through correspondence, and anagrams. Beginning in the seventeenth century, learned societies and their journals constituted new mechanisms for establishing priority. The famous priority dispute between Newton and Leibniz over the discovery of the calculus illustrates that institutions themselves—the Royal Society of London in this case—could serve as formal (if not necessarily impartial) tribunals to adjudicate claims.[2] With the advent of the scientific societies researchers could also document their priority by depositing sealed envelopes with institutional secretariats. The Academy of Sciences in Paris has served as such a depository from its orgins in the seventeenth century.[3] However, the main mechanism that arose in the seventeenth century for establishing priority was dated publication in learned-society periodicals, and, indeed, according to A. J. Meadows, "The urge to establish priority formed one of the essential factors in the initial establishment of scientific journals."[4]

The Comité de Librairie was an early and key agency performing these historically new functions establishing and enforcing priority and rights over scientific work.[5] The Comité discharged these functions rigorously as a regular part of its activities, primarily by keeping track of the date materials were read to the Academy and/or presented to the Comité. The registers list items considered by the Comité by date, and beginning in 1770 the secretary, Grandjean de Fouchy, introduced a new rubric, "Dates," to record official dates for papers.[6] In some instances academicians announced papers to the Comité in order to establish a date, as was the case, for example, for P.-J. Macquer's piece on an emetic, which elicited the response from the Comité: "It is only an announcement. Wait for the memoir. Good only for the date."[7]

By and large the Comité respected the dates of papers and published them in the order in which they were inscribed on its registers. A young P.-S. Laplace seems not to have appreciated this fact when he asked the Comité to advance the publication of one of his papers. It refused. Laplace repeated his petition, arguing that the piece in question extended earlier work which had already appeared in the *Mémoires*. This time, the Comité responded explicitly that only memoirs "submitted to the secretariat and initialed by the secretary during the year" would be inserted in the annual volume in question.[8] And so it was that papers ordinarily appeared in the *Mémoires* with an official date noted in the margin (Fig. 6).

Confusion over dates could and did arise, especially because authors had opportunities to revise papers after they were read to the Academy or approved by the Comité. Compounding this problem, a volume of the *Histoire et Mémoires* typically appeared several years after the presentation of papers, a notorious time lag that tended to lead authors to make changes in order to keep up with intervening work.[9] Condorcet brought this problem to the Comité's explicit attention in 1776:

> [He] observed that among the memoirs for the 1773 volume there were several which had just been remitted to the secretariat and others which are still in the hands of authors. The Comité ruled that it would publish these memoirs with the marginal note: *submitted by the author on . . .* [sic]. A memoir cannot have any other authentic date.[10]

The concern that published work should exactly match what was read to the Academy extended to works published outside the Academy by nonacademicians. Complaints arose, for instance, that A.-A. Parmentier introduced changes in the published version of his *Expériences et refléxions*

DES SCIENCES. 185

EXPÉRIENCES

SUR LA

RESPIRATION DES ANIMAUX,

Et sur les changemens qui arrivent à l'air en passant par leur poumon.

Par M. LAVOISIER.

DE tous les phénomènes de l'économie animale, il n'en est pas de plus frappant ni de plus digne de l'atten- tion des Physiciens & des Physiologistes , que ceux qui accompagnent la respiration. Si d'un côté nous connoissons peu l'objet de cette fonction singulière , nous savons d'un autre qu'elle est si essentielle à la vie , qu'elle ne peut être quelque temps suspendue, sans exposer l'animal au danger d'une mort prochaine.

    L'air , comme tout le monde sait , est l'agent , ou plus exactement le sujet de la respiration ; mais en même-temps toutes sortes d'air , ou plus généralement toutes sortes de fluides élastiques, ne sont pas propres à l'entretenir , & il est un grand nombre d'airs que les animaux ne peuvent respirer sans périr aussi promptement au moins que s'ils ne respiroient point du tout.

marginal note: 3 Mai 1777.

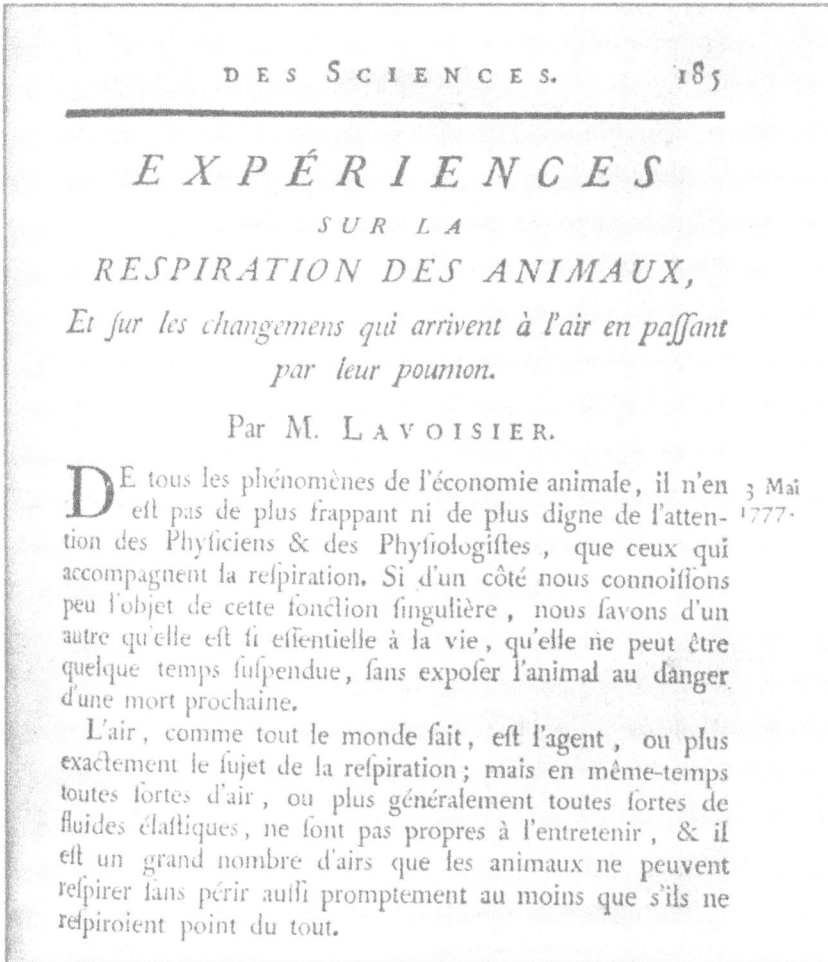

Figure 6: Lavoisier's memoir on animal respiration with marginal date indicating when the paper was read to the Academy. MARS 1777 (1780), p. 185.

relatives à l'analyse du bled et des farines (1776) after he read it to the Academy, and the Comité appointed commissioners to investigate the potential ruse.[11]

    In a poignant instance the rules concerning dates and the publication status of authors conflicted, much to the detriment of Jean-Baptiste Le Roy. In 1748 he and count Patrick d'Arcy submitted a memoir that they had written together on measuring electricity. The next year the Academy named d'Arcy an adjunct in mechanics, at which point d'Arcy re-read the paper to the Academy. Because only academicians were entitled to publish in the *Mémoires*, the Comité scheduled the paper to appear under d'Arcy's name alone, a fact Le Roy protested to the Comité, which then ordered the following:

> Because Mr. Le Roy became a member of the Academy in 1751 it would be impossible to put his name at the head of a paper of 1749, but the Comité would place an asterisk after Mr. d'Arcy's name and add the following footnote: *This memoir is printed under the name of Mr. d'Arcy alone although it belongs equally to Mr. Le Roy because when Mr. d'Arcy read this memoir Mr. Le Roy was not yet a member of the Academy.*[12]

And so the paper appeared with justice served.

In some instances, the perceived interests of science and the Academy dictated that a paper appear before its appointed time, a practice known as "anticipation." The most famous example concerned the observations of the 1769 transit of Venus, which were rushed into print in 1770 so that astronomers throughout the Republic of Letters might make use of them as soon as possible.[13] Regarding Tillet's piece of 1762 on an insect infestation of the wheat crop in the Angoumois, the Comité decided to "publish in [the volume for 1761 in press] with a note giving the reason for this anticipation."[14] Of the Marquis de Vallière's 1775 memoir on "the superiority of long and solid artillery pieces over short and light ones," the Comité voted to "print [it] in the volume for 1772 [in press], given the circumstances and importance of the subject," these presumably arising from its inherent utility or impending war with England.[15] Another motive for speeding up the publication of papers was to have pieces on similar topics appear together. Guettard's 1755 piece on the mineralogy of Canada, for example, appeared in the 1752 volume to accompany a piece on the mineralogy of Switzerland.[16] Along these lines, the Comité once agreed to include in the *Mémoires* observations of a solar eclipse from nonacademicians, but "following those of the academicians."[17] Academicians so routinely petitioned the Comité to push up the publication dates of papers that in 1777 Condorcet added another category to the Comité's registers, "Demandes particulières." Although a consistent rationale for granting approval for early publication is not apparent in the record, permission was often predicated on space being available in the volume in question.[18]

The Comité's last-minute decision to insert a memoir from the Montpellier Society of Sciences in the Paris volume for 1746 had unfortunate, but revealing consequences. The memoir in question, the abbé F.-B. de Sauvages' first paper on the geology of the Languedoc, was read in Paris in 1749 and 1750, and by rights it should have appeared in the volumes for those years. But the need to place a Montpellier piece dictated that it appear in the volume for 1746, then in press. A related mineralogical map by the academician Guettard was also slated for the 1746 volume. Guettard had read his memoir in 1746, but the simultaneous publication of these two papers might seem to privilege Sauvages' later work over Guettard's earlier piece. That Sauvages referenced a paper of his published in 1743 added to the appearance of his priority over Guettard. At a special meeting, the Comité de Librairie voted to add a note to the Sauvage piece giving the official dates for his and Guettard's papers: "We report all these dates exactly in order to avoid confusion that might result from having advanced the publication of some memoirs, and it is [only] fair to preserve exactly the dates for those to whom they belong." (Be it noted that when the Comité consented to publish Sauvages' second memoir in the volume for 1747, it "resolved to send the proof-sheets to Mr. Guettard"!)[19]

The Comité occasionally speeded up publication to establish the priority of its own members. Associate in mechanics, the marquis de Courtivron asserted claims over the Prussian academician, Samuel Koenig, in the discovery of the principle of mechanics that links statics and dynamics. Courtivron petitioned the Comité "that the title of a memoir he read in 1749 on a new principle of dynamics appear in [the volume for] 1747 with a note in the history section, given that Mr. Koenig has already used the same principle and is presently publishing a work that employs it."[20] The Comité concurred, and, although the requested notice did appear in the volume for 1747, it was omitted from the table of contents. Courtivron complained that, because "the printer had omitted the title from the table of contents and, more, had made a major mistake in announcing the principle, [Courtivron] consequently asked that the same title and the same announcement be inserted in the volume for 1748," which the Comité again approved.[21] Comparison of the two announcements is revealing. The first for 1747 begins: "The memoir whose title is given here will be printed in the Academy's volume for 1749. Particular reasons have elicited this announcement of the general principle of which one will find the demonstration and application in the memoir when it is published."[22] The 1747 announcement goes on to (mis)state the principle that the position of static equilibrium of a connected system of bodies is the same as that when, the system under motion, the product of the masses of the bodies and their speeds (mv) reaches a maximum. The 1748 version is verbatim as above, but continues: "The title having been omitted in the table of contents for the 1747 volume, where it should have appeared, we provide it in this one, and we correct a printer's error over-looked in the announcement of this principle."[23] This time the principle is correctly expressed as the product of the masses and the *square* of their speeds (mv^2). Courtivron's paper appears as promised in the volume for 1749. The effort may have been laudable to certify the priority of one of their own, but one can question whether the Comité erred in Courtivron's favor and whether Courtivron's initial formulation of 1747 was indeed a "printer's error."[24]

Notes

1. Price, *Big Science*, esp. chapt. 3, provides the standard account.

2. Hall, *Philosophers at War*.

3. On "plis cachetés" in the Academy, see Brian and Demeulenaere-Douyère, *Histoire et mémoire*, 73–75.

4. *Communication in Science*, 55; see also 55–60, 67. Meadows goes on to note that the success of the early scientific journals as vehicles for establishing priority claims of authors largely eliminated secrecy from the world of scientific investigators, with the printed book likewise declining in importance as a mechanism for authors to assert claims over their scientific discoveries.

5. It is notable in this connection that priority disputes simply did not arise at the Académie des Inscriptions et Belles-Lettres. The instance that proves the rule concerns a request by the abbé Barthelemy "de faire imprimer dans le Journal des Savans un extrait de son Memoire sur quelques Medailles Samaritaines pour empecher que quelque Savan étranger, entre les mains du quel pourroit tomber une des gravures de ces Medailles qu'on a distribuées au public ne lui enlevât le merite de cette decouverte"; Institut de France, Archives de l'Institut, Cote A 71, "Registre des Assemblées & Délibérations de l'Académie Royale des Inscriptions et Belles Lettres pendant l'Année 1790," meeting for Tuesday, 20 April 1970. At root is the difference in production and rewards in the sciences and the humanities.

6. RCL II/9 (December 1770).

7. RCL I/100 (December 1760): "ce n'est qu'une annonce attendre Le mémoire bon seulement pour La Date."

8. RCL II/64 (December 1776): RCL II/65 (February 1777): "Il a été decidé qu[e le Comité] n'inserverait pas pour 1773 ou 1774 que les memoires remis au Secretariat pendant ces années et paraphès par le Secrétaire." See earlier concern expressed by the Academy as a whole that dates be respected; PV 70 (1751), p. 280.

9. The best known instance of this kind of change is Lavoisier's famous "Easter Memoir" of 1775. In a paper read at the Easter meeting of the Academy in 1775, Lavoisier announced his discovery that ordinary air is taken up in combus-

tion. In the published version in the *Mémoires* three years later, he had reformulated his views to the effect that "eminently respirable air" (oxygen), not common air, became absorbed in combustion, a nontrivial change emphasized by Conant in his "The Overthrow of Phlogiston Theory."

10. RCL II/64 (December 1776): "J'ai observé que parmi les Memoires pour 1773 il y en avait plusieurs qui venaient seulement d'estre remis au sécretariat, et d'autres qui sont encore entre les mains des auteurs. il a eté statué que L'on mettrait à La marge de ces mémoires *remis par L'auteur le*. . . . Le mémoire ne pouvant pas avoir une autre date autentique."

11. The commissioners, L.-C. Cadet and N. Desmarest, exonerated Parmentier; see RCL II/59 (March 1776) and PV 95 (1776), 114.

12. RCL I/24 (May 1752): ". . . Mr. Le roy n'ayant été nommé de L'acad. qu'en 1751 il etoit impossible de mettre son nom en tête d'un mémoire de 1749 mais qu'on mettroit en renvoy après le nom de mr. Darcy avec cette note au bas de la page: *Ce mémoire est imprimé sous le nom de mr. Darcy seul quoy qu'il appartienne également à Mr. Le roy parce que lors que Mr. Darcy a leu ce mémoire Mr. Le Roy n'étoit pas encore de L'académie.*" See the published note MARS (1749), 63n, where the wording differs slightly.

13. See MARS (1767), 643. For similar anticipations regarding the 1761 Venus transit, see RCL I/110–11 (February 1762) and HARS (1757), 77ff. "To satisfy the demands of astronomers," a comparable effort speeded into print the results of Lacaille's astronomical observations from the Cape of Good Hope in 1751–1752; see MARS (1748), 601–612 and note on 612; RCL I/47 (November 1754); and MARS (1751), 456n, 519n.

14. RCL I/114 (May, 1762): "Imprimé en 1761 avec une note qui donne la raison de cette anticipation." See also MARS (1761), 289–331, for the paper by Duhamel and Tillet. The *Histoire* section for 1761 [HARS (1761), 66] mentions that Turgot sent these academicians on a fact-finding mission to the region, the evident public utility of their findings accounting for the "anticipation."

15. RCL II/53 (August 1775): "Imprimé en 1772 attendu La circonstance et Limportance de la matière." See also HARS (1772), pt. 2, 44–55; and MARS (1772), pt. 2, 77–114.

16. RCL I/46 (July 1755), I/47 (November 1755). See also RCL I/55 (May 1756), where the Comité moved up Guettard's 1755 article on stalactites to accompany Daubenton's on alabaster, "mais après celuy de Mr. D'aubenton."

17. RCL I/147(August 1765): "à la suite de celles des académiciens."

18. See, for example, RCL II/39 (March 1774); II/69 (July 1777).

19. On this episode, see RCL I/10 (July 1750); I/15 (March 1751); and MARS (1746), 713–758 and notes on 713, 714, and 758, for the quotation. See also RCL I/18 (July 1751): "il a été resolu que les épreuves seroient renvoyées à Mr. Guettard."

20. RCL I/19 (August 1751): "il demande que le titre d'un mémoire qu'il a leu en 1749 sur un nouvel principle de dynamique paroisse en 1747 avec une note dans Lhistoire attendu que Mr. Koenig a dejà fait usage du même principe et publie actuellement un ouvrage où il est employé ce que le Comité a accordé."

21. RCL I/23 (March 1752): "L'Imprimeur avoit omis ce titre dans la Table et de plus avoit fait une faute considérable dans l'énoncé du principe et demander en conséquence que le même titre et le même énoncé soient inserés dans le Volume de 1748 ce que Le Comité luy a accordé."

22. MARS (1747), 698.

23. MARS (1748), 304.

24. See also HARS (1749), 177–79 and MARS (1749), 15–28.

⦚ Chapter Six ⦚

Monitoring Disputes

*T*he Academy of Sciences and its Comité de Librairie provided the primary mechanism for establishing priority in the world of eighteenth-century French science. As seen in Chapter 5, they exercised this role by policing the dates that academicians and others submitted their work to the institution. But the close attention devoted to dates did not prevent conflicts from breaking out. In fact, inside and outside the Academy priority disputes formed such a part of eighteenth-century French scientific life that contemporaries had a special word for them: the "contestation," a term with overtones of an intellectual duel.[1] This chapter details how the Comité de Librairie, meeting behind closed doors, functioned as a scientific tribunal that adjudicated quarrels that arose between academicians.[2]

On the simplest level and in line with the Academy's statutes, the Comité de Librairie insisted on decorum and avoiding personal attacks in print. Article xxvi of the Letters Patent of 1699, for example, required that "the Academy will carefully see to it that, when academicians are of different opinions, they will not use any harsh or scornful language against one another either orally or in writing, and when they oppose the views of anyone in the learned world, the Academy will demand that they do so only with moderation."[3] Along these lines the full Academy once formally reprimanded C.-M. de La Condamine for attacking J.-E. Guettard personally in the public press and for thus casting the Academy in a bad light.[4] In general, however, it fell to the Comité de Librairie to monitor the tone of materials destined for publication. Regarding the same Guettard's paper on the specific character of shellfish, for instance, the Comité ruled: "Print, but cut out everything concerning Mr. Dargenville."[5] It likewise rejected a piece by Duhamel because it cast aspersions on the work of Vicq d'Azyr.[6]

Overall, academicians were jealous of one another and kept a suspicious eye out for what their colleagues might sneak into print. On one occasion, P.-C. Le Monnier added a footnote to a paper that implied that his work was the stimulus for an Academy prize contest. When the matter came before it, "The Comité decided to suppress [Le Monnier's] note and paste a label over it."[7] In another case, the secretary reported to the Comité that "several people remarked that the end of Mr. Daubenton's memoir on precious stones might not conform to the views of the Academy." After hearing Duhamel's report, the Comité found the piece innocuous and "good for publication."[8] These minor examples establish a baseline for the more notorious disputes that the Comité and the Academy came to handle.

As a scientific tribunal, the Comité kept track of claims, tried to facilitate communications between warring parties, and at times appointed commissioners to investigate matters. Three cases that significantly occupied the Comité de Librairie illustrate its role in dealing with priority disputes.

Bones of Contention

J.-R. Tenon and F.-D. Hérissant became embroiled in a priority conflict that seems typical, if depressingly trivial in its details. In April 1758 Hérissant, an associate anatomist, presented a paper on bone growth, which the Comité approved for the 1758 volume of the *Mémoires*. In December 1758 Tenon, not yet a member of the Academy, began reading his own paper on bones. The Academy named Hérissant and Duhamel to examine Tenon's paper, and in June 1759 they recommended it for the *Savants Étrangers* series. Tenon was then elected an adjunct anatomist, so his paper was returned to him, and he re-read it and a second paper on the same subject at sessions of the Academy in July and August of 1759, "Mr. Hérissant being present and making no objection." At a public meeting in November 1759, Hérissant read another memoir on bones, but he delayed submitting this paper until 1762. At that point, he asked that his paper appear in the volume for 1759, the year he first presented it. (The volume for 1759 did not appear until 1765.) Tenon took exception to this maneuver, claiming that in the 1762 version of his 1759 paper Hérissant used Tenon's experimental methods without acknowledgment, and thus publication of Hérissant's paper in the volume for 1759 would make it appear as if Tenon did not originate these approaches to studying bone growth and disease. The Comité sided with

Tenon, ruling that Hérissant's 1759 paper could only appear in the volume for 1762, "it not being possible to say that the content of the memoir read in 1762 is substantially the same as that read in 1759." However, because the Comité feared that the projected volume for *1758* was not large enough, it proceeded to add Tenon's papers of 1758 and 1759. (That Tenon was not a member of the Academy in 1758 and technically not entitled to publish in the *Mémoires* for that year was felt to be moot, given that the papers had remained secure in the secretariat.) Hérissant naturally protested and asked that his paper of 1759/1762 likewise be included in the 1758 volume, to which Tenon objected in turn. The Comité ultimately ruled that all the papers would appear together in the volume for 1758, Tenon's with the priority dates of December 1758 and August 1759, and Hérissant's with the dates of April 1758 and December 1762.[9] For Tenon, the appearance of forfeiting priority was a serious concern. For us, the mechanisms at play to resolve his anxieties are of greater historical interest.

Bouguer and La Condamine Fight It Out

The battle that erupted between Pierre Bouguer and Charles-Marie de La Condamine was more vicious and occupied the Academy and the Comité de Librairie to a more significant extent. Like the conflict between Tenon and Hérissant, at issue was not a point of scientific disagreement, but simply who was to get priority credit. At stake in this case was credit, not for minor work, but for contributing to the most important scientific research done in the first half of the eighteenth century: the Academy-sponsored efforts of the 1730s and 1740s to measure a degree of arc at the pole and at the equator and to decide between the competing world views of Newton and Descartes. Newton had predicted that the earth was flattened at the poles, whereas Cartesian science said that the spin of the ethereal vortex caused the earth to flatten along the equator. The Academy famously undertook to decide between these two predictions by sending expeditions to Lapland and to Peru to measure the earth's curvature.[10] Bouguer and La Condamine, along with Louis Godin, headed the scientific party that journeyed to South America. A compounding element of the dispute that subsequently arose between Bouguer and La Condamine thus derives from the fact that they had started out as collaborators.

Three years older than La Condamine, Bouguer entered the Academy a year after his adversary, but rose more quickly to the rank of pensionnaire. The two did not get along during the expedition to the equator and had feuded. Bouguer returned to France first, in 1744, and read papers before the Academy under his name alone that stemmed from his collaborative work with La Condamine in South America. When La Condamine finally got back to Paris in 1745, he protested that fact and Bouguer's stubborn refusal to communicate to his former partner what he had already presented to the Academy. La Condamine lamented at one point: "Mr. B[ouguer] has taken possession of everything in presenting our common work first . . . it is impossible for me to say anything new."[11]

Bouguer beat La Condamine into print with his "Abridged Account of the Voyage to Peru" that appeared in the *Mémoires* for 1744 published in 1748.[12] Bouguer shamelessly sought to capitalize on his initial advantage by refusing to have his papers copied into the Academy's registers, as the statutes required, arguing that La Condamine could thereby access his work without Bouguer reciprocally knowing what La Condamine might report. Only because La Condamine threatened to make the affair public was his formal protest entered into the Academy's record.[13] The conflict became so heated that in 1749 the Academy's president, Jean-Jacques Amelot, met privately with the two academicians and worked out a compromise whereby another paper by Bouguer would appear in the volume for 1746 along with extracts that La Condamine would be allowed to prepare.[14] However, when La Condamine tried to read his extract before the Academy, Bouguer interrupted at every turn to stall matters, possibly because the Comité had granted La Condamine only a limited time to submit his observations.[15]

Bouguer's and La Condamine's papers appeared together in the 1746 *Mémoires* published in 1751, but without any mention in the *Histoire* section or any indication of the discord surrounding them.[16] But matters did not end there. Bouguer went on to complain that La Condamine had introduced changes in the published version from the manuscript original copied into the Academy's minutes. Appointed to investigate, P.-C. Le Monnier found the differences "real," whereupon the Comité voted to "exhort Mr. de La Condamine to deduce the reasons he could have had to make these changes."[17] A superficial examination reveals only minor, largely stylistic, differences between the manuscript and published versions, but regardless of their scientific significance, Bouguer used them as the basis to demand that the Academy insert a notice in the volume in question, "declaring that it cannot in any way pretend to certify the facts advanced by Mr. de La Condamine." It was too late to do anything for the 1746 volume, but the Comité offered to "see what can be done for the one for 1747."[18]

Controversy flared up again over related papers in 1751 and 1753, and also when it came time for the two men to publish their complete results as books. It is not necessary to enter into those sad details to recognize that the Academy and the Comité de Librairie functioned as behind-the-scenes tribunals where often bitter quarrels between academicians were contested.

Static between Nollet and Le Roy

Other conflicts more directly concerned scientific disagreements. The premier example is the dispute that arose between the abbé Jean-Antoine Nollet (1700–1770) and Jean-Baptiste Le Roy (1720–1800) over electrical theory. In the mid-eighteenth century no area of science more captured the imagination of the educated public or the attention of scientific investigators than did effects associated with (static) electricity.[19] The invention of machines and techniques to generate, transmit, and store electricity at once allowed for popular public entertainments and for scientific inquiry into the phenomena, and from the 1730s the French led in both.

Tutor to the royal children, professor at the Collège de Navarre, academician in 1739, and author of the oft-reprinted *Leçons de physique expérimentale* (6 volumes, 1743–1748), the abbé Nollet was the "foremost electrical theorist of the mid-century."[20] He entered the lists as a theorist in 1745 with a paper, "Conjectures on the Causes of Electricity in Bodies," published in the *Mémoires* of the Academy for that year, and he continued to publish on electricity and to defend his views in the *Mémoires* and elsewhere through the later 1760s.[21] Nollet argued for the existence of a single electrical fluid, whose various inward and outward flows accounted for the observed phenomena.[22] At mid-century an almost universal consensus emerged in the community of French physical scientists to support Nollet's one-fluid theory.

Twenty years younger than Nollet, Le Roy was elected to the Academy as a junior, adjunct member in 1751, and in the period of concern here he was decidedly the young turk, not receiving promotion to the ranks of Academy associates until 1766. Le Roy was Nollet's protégé, but he turned on his mentor, offering a two-fluid theory of electricity in papers read to the Academy in 1753 and 1755.[23]

Nollet's and Le Roy's theories embodied contradictory and mutually exclusive interpretations of the phenomena of electricity, but more was involved than the nominal science. Part of the conflict was generational,

with Nollet using his position as the senior and more established academician to thwart Le Roy and other theorists for the better part of two decades.
Also at issue were nationalistic concerns that pitted Nollet and the strong
tradition of French electrical research against Benjamin Franklin and the
upstart electrical theory emanating from the Americas. Finally, and perhaps most importantly in terms of understanding the onset of the episode,
the Nollet–Le Roy imbroglio was a surrogate for a larger factional struggle
in the Academy between Réaumur and Buffon. Buffon and Réaumur were
enemies, and Nollet was Réaumur's protégé, so when Franklin's
Experiments and Observations on Electricity was first published in England in
1751, Buffon quickly had the work translated into French as a way to
foster opposition to Nollet and his views of electricity. The young Le Roy
entered the fray as a supporter of Franklin, a member of the Buffon clique,
and hence, in Nollet's eyes, an ingrate and an apostate.

John Heilbron, who documents this story, underscores that Nollet's
theoretical explanations for electricity began to face serious threats early
in the 1750s. Part of the danger stemmed from the difficulties Nollet had
in providing a theoretical account of the Leyden jar, the apparatus (a con-
denser) for storing electricity that had been invented after Nollet's initial
theoretical formulations of 1745. The other part of the peril derived
from Franklin's alternative views of the nature of electricity that were
being pushed by Le Roy. The theoretical and experimental points in con-
tention were esoteric and nonobvious, and Franklin's views were not without their problems, either. In the face of challenges to his scientific
authority, Nollet went on the attack. When Le Roy read and published
papers in support of the existence of two forms of electricity, Nollet got
the Comité de Librairie to agree to publish his refutations alongside Le
Roy's papers.[24] In 1752 and again in 1760 he asked for and got the
Academy to appoint commissioners to certify certain "electrical facts"
that supported his one-fluid position. In both instances the commissioners, which in 1752 included Le Roy, strongly sided with Nollet, no doubt
because, in what Heilbron labels a "swindle," Nollet wrote the reports.[25]

The two sides agreed that a similar commission should investigate
Le Roy's claims, which a six-man investigating team did in early 1762.[26]
Le Roy's work came up for review in the Academy on Wednesday,
February 24, 1762. Nollet himself, as the director for that year, summa-
rized the dispute, and the commissioners then reported on nine of Le
Roy's electrical experiments involving glass and sulphur globes, conclud-
ing that "such are the facts we have witnessed and about which we are con-
fident."[27] The Academy then voted to ask its commissioners for their

opinion regarding the claims of the respective partisans. The group deferred until the next meeting, when it addressed four written propositions concerning Le Roy's scientific claims. In each case the committee sided against Le Roy: electricity from sulphur globes does not equal that drawn from glass globes, sparks emanating from charged rods are always "tufted," visible sparks do not signal the "entry" of electrical fluid, and a conductor does not remain neutral when suspended between charged globes of sulphur and of glass.[28]

At this point the battle shifted to the Comité de Librairie where Le Roy asked that his memoir "on the Leyden jar and the exposition of the facts that can help us learn their causes" be published in the *Mémoires*. He went on to say:

> If one raises the commissioners' report as an objection, I respond that one only has to print the report at the end [of my paper] along with my further observations on what it contains. It seems to me that I cannot justly be denied this. Every academician is responsible before the public for what he proposes. The public will judge my memoir. It will judge the commissioners' report. It will judge my remarks on the report's contents.[29]

Nollet had no objection to this procedure, provided that he, too, had an opportunity to publish his own response. The Comité de Librairie, however, rejected these alternatives. Rather, it decreed that Le Roy's paper could appear only if "he suppress or at least correct" two propositions contradicted by the report. The first concerned the "tufted" character of sparks issuing from (and not entering) charged rods. The second represented a more serious matter of scientific politics, for Le Roy was claiming "that Mr. Franklin was the first to begin to uncover the mysteries" of the Leyden jar. To this claim the Comité responded sharply:

> This assertion does not accord with the history of the Academy which is proud that, beginning in 1746—more than seven years before anyone heard Mr. Franklin mentioned in France—Messrs. Nollet and doctor [Louis-Guillaume] Le Monnier had not only repeated and analyzed the Leyden experiment but, more, indicated the causes and the reasons for it, which were printed in the volume for 1746. From which it follows that Mr. Le Roy today cannot strip these two academicians of the priority that belongs to them in order to vest it in a foreigner who seems not to merit any right to it.[30]

(Ironically, this report ignores the fact that Nollet feuded with Le Monnier over their electrical papers in 1746![31]) Finally, Le Roy's idea that "every academician is responsible before the public for what he proposes" or that the public should decide these matters struck at the heart of what the Academy and the Comité were charged to do: "examine and judge" science for the public and for the community of scientific investigators. Le Roy's suggestion regarding the public perhaps bespoke the faint, but rising chorus that the public was an entity with a voice in public matters in the Old Regime. In any event, in response to Le Roy's proposal, the Comité continued in an even harsher vein:

> Finally, the expedient that Mr. Le Roy proposes to print the memoir along with the report cannot be accepted. It would be too strange to see in the Academy's very own volume one of its members appealing its judgment, so to speak, to the public and to somehow render false the facts established by six of its members named for that purpose.[32]

The Comité de Librairie wanted it understood that the public did not have the right to judge and certify truth claims about the natural world. That right belonged to the Academy and to the Comité.

Better attuned to these considerations, Nollet then offered his memoir for publication, after purging "everything concerning Mr. Le Roy and their dispute and communicating the memoir to any commissioner the Comité deemed appropriate."[33] The Comité accepted this offer and charged Duhamel to review Nollet's manuscript. Duhamel reported favorably, and the paper by the victorious Nollet, "Reflections on some Phenomena cited in Favor of Positive and Negative Electricity," appeared alone in the volume of *Mémoires* for 1762.[34]

The historiography of the eighteenth-century Academy has made clear that the institution judged the science produced by outsiders. Its rejection of Marat and Mesmer as scientific charlatans are famous instances. However, regarding the work of its own members or other accredited researchers, the assumption seemingly has been of an even-handedness and a reluctance to play a partisan role. The Nollet–Le Roy case, to the contrary, reveals that within the institution the Academy and the Comité de Librairie were not neutral parties in these disputes, but responded to intricate interplays of institutional and national interests, internal politics, seniority, tradition, and patterns of scientific allegiance. John Heilbron's conclusion is that Nollet "bamboozled" the Academy and that the Nollet–Le Roy feud "paralyzed" the institution, divided it along generational lines, and effectively stymied French electrical science until after Nollet's death in 1770.[35] The partiality of the Academy and the Comité de Librairie toward Nollet highlights an unsurprising sense that the Comité represented the old guard within the Academy using its powers to preserve the commitments of its members and so to shape the content of the science produced under its auspices. Perhaps Le Roy felt vindicated when he was finally elected a *pensionnaire* of the Academy in 1770 to fill the seat made vacant by Nollet's demise.

Notes

1. See, for example, RCL I/67 (August 1757), where the record speaks of "une Contestation qui s'éleva entre M. Darcy and [M. Deparcieux]."

2. It is worth noting in this connection that the Académie Royale des Inscriptions et Belles-Lettres refused to become involved when disputes arose between members of that academy; see Institut de France, Archives de l'Institut, Cote A 67, "Registres des Assemblées & Délibérations de l'Académie Royale des Inscriptions et Belles-Lettres," meeting for Friday, 9 February 1781, where, regarding a dispute between the Marquis de Paulmy and one M. Le Grand, the academy decided "qu'elle persiste dans la resolution déjà prise de n'être ni juge, ni arbitre."

3. In a like vein Article XXXVII directed the Academy's annual president "to be very attentive that good order be maintained in the meetings and in everything affecting the Academy," and it required him to report directly to the king or the minister in charge of the Academy about these matters. Articles XXVI and XXXVII of the Letters Patent of 1699 are reprinted in Brian and Demeulenaere-Douyère, *Histoire et mémoire*, 411–12.

4. PV 84 (1765), fols. 403v–404v.

5. RCL I/55 (May 1756): "Imprimé en retranchant tout ce qui Concerne M. Dargenville."

6. RCL I/62 (August 1776).

7. RCL I/9 (April 1750): "Le Comité a décidé que cette note seroit supprimée et qu'on mettroit un carton à cette feuille." The minutes go on to remark: "Cecy n'a point eu D'exécution." Le Monnier appealed the decision of the Comité to the full Academy which may account for the lack of follow-through in this matter; see PV 69 (1750), 161, meeting of Wednesday, May 6, 1750.

8. RCL I/12 (December 1750): ". . . quelques personnes avoient trouvé que la fin du mémoire de mr. Daubenton sur les pierres precieuses pouvoir n'estre pas conforme au sentiment de L'academie . . . Vû le rapport bon a imprimer." See also Daubenton's paper, MARS (1750), 28–38, esp. 37. At issue was Daubenton's claim that naturalists' classifications of species are arbitrary. Daubenton avoided censorship because he granted that true species do exist in nature.

9. RCL I/126–128 (July 1763): "sans que Mr. herrissant qui y etoit present fût aucune réclamation . . . n'etant pas possible de justifier que le mémoire lû en 1762 est absolument le même quant au fond que celuy qui a été Lu Le 14 novembre 1759." See also lengthy complaint by Tenon in RCL I/120–121 (January 1763), and published papers in MARS (1758). In print, Tenon's first paper carried the marginal date of July 20, 1759, but with a footnote stating that "M. Tenon read this memoir on December 6, 1758 before becoming a member of the Academy."

10. Gillispie, *Science and Polity*, 112–113; McClellan, *Science Reorganized*, 200–202. For a detailed account of the science involved in these expeditions and their results, see John L. Greenberg, *The Problem of the Earth's Shape*.

11. PV 69 (1750), 323–328: "Mr. B. . . . s'est mis en possession de tout ce qu'il a dit le premier, sur nôtre travail commun . . . il m'est comme impossible de rien dire de nouveau."

12. "Relation abrégée du voyage fait au Pérou par Messieurs de l'Académie Royale des Sciences, pour mesurer les degrés du méridien aux environs de l'équateur, et en conclure la figure de la Terre," MARS (1744), 249–297.

13. This skirmish unfolded in November and December of 1748, PV 67 (1748), 489–491, 499, and 519–520. Article XXIV of the Letters Patent of 1699 required that papers be deposited with the secretary the day of their reading; Brian and Demeulenaere, *Histoire et mémoire*, 411.

14. PV 68 (1749), 188.

15. RCL I/6, 7 (February 1750); see also La Condamine's protest, PV 69 (1750), 232–234.

16. Bouguer, "Suite de la relation abrégée donnée en 1744, du voyage fait au Pérou pour la mesure de la Terre," MARS (1746), 569–606. La Condamine, "Extrait des opérations trigonométriques, et des observations astronomiques, faites pour la mesure des dégres du méridien aux environs de l'équateur," MARS (1746), pp. 618–688.

17. RCL I/16 (April 1751): ". . . il a été décidé que je . . . L'exhorterois à deduire les Raisons qu'il peut avoir euës de faire ces changements."

18. RCL I/18 (July 1751): ". . . L'acad. Declairait qu'elle ne pretend estre en aucune facon Garante des faits avancés par mr. DelaCondamine sur quoi il a été decidé qu'il ne pouvoit se faire aucun changement au Volume de 1746 et que pour Voir ce qui pouvait avoir lieu dans celuy de 1747." Nothing seems to have resulted from Bouguer's request.

19. For background, see Heilbron, *Electricity*, esp. 352–362.

20. Gillispie, *Science and Polity*, p. 506.

21. "Conjectures sur les causes de l'électricité des corps," MARS (1745), pp. 107–151. See list of Nollet's publications in Heilbron, *Electricity*, 548–549.

22. Heilbron, *Electricity*, 280–289, discusses the details of Nollet's system.

23. "Mémoire sur l'électricité," MARS (1753), 447–474; "Mémoire sur l'électricité résineuse," MARS (1755), 264–283.

24. See Nollet's "Examen de deux questions concernant l'électricité," MARS (1753), 475–515, and his "Suite du mémoire dans lequel j'ai entrepris d'examiner si l'on est bien fondé à distinguer des électricités en plus et en moins, résineuse et vitrée," MARS (1755), 293–317. See companion papers by Le Roy cited in previous note and dealings in the Comité de Librairie regarding these papers, RCL I/48 (November 1755), RCL I/56 (June 1756), RCL I/59 (August 1756), RCL I/71 (February 1758), the latter entry reading, "M. L. Nollet a demandé qu'un mémoire Intitulé Mémoire Contre la Double électricité Résineuse et Vitrée Lû le 28 Aoust 1756 et Destiné à L'impression lequel soit de réponse à celuy que M. Le Roy a leu en 1755 fût imprimé dans ce même Volume avec celuy de M. Le roy. Ce qui lui a été accordé." The volume of the *Mémoires* for 1755 did not appear until 1761.

25. Heilbron, *Electricity*, 359. See also RCL I/26 (August 1752); PV 71 (1752), 482, 548–556; PS, 23 August 1752; and report, PV 79 (1760), 240–252.

26. PV 79 (1760), 358v; PV 81 (1762), 3.

27. PV 81 (1762), fols. 73–73v: "Tels sont les faits dont nous avons été témoins et dont nous nous sommes assurés. . . ."

28. PV 81 (1762), 74–81, meeting for Saturday, February 27. Nollet had declared the latter a decisive experiment distinguishing his interpretation from Le Roy's.

29. RCL I/112–113 (April 1762): "Si L'on m'objecte le rapport de Mrs. les Commissaires Je répond qu'on n'a qu'à LImprimer à La suite avec les Observations que J'ai faites sur ce qu'il Contient observations que Je suis prêt à Lire. il me semble qu'on ne peut me refuser cette Justice. Tout Académicien est responsable devant le public de ce qu'il avance il jugera mon mémoire il jugera le rapport de Mrs. Les Commissaires il Jugera les Observations que J'ai faites sur ce qu'il Renferme."

30. RCL I/112–113 (Avril 1762): "Cette assertion ne peut s'accorder avec L'histoire de L'Académie qui fait foy que dès 1746 plus de 7 ans avant qu'on eut entendu parler en france de M. franklin, Messieurs L'abbé Nollet et le Monnier medecin avoient non seulement répété et analysé L'expérience de Leyde mais encore en avoient indiqué les Causes et les raisons qui ont été imprimées dans le même Volume d'où il suit que M. Le roy ne peut aujourd'huy dépouiller ces deux académiciens d'une primauté qui leur appartient pour en revêtir un étranger qui ne paroît y avoir aucun droit. . . ."

31. See RCL I/13 (December 1750).

32. RCL I/112–113 (April 1762): ". . . enfin L'expedient que propose M. Le Roy d'Imprimer le mémoire tel qu'il est avec le rapport ne peut estre admis il seroit

trop singulier de Voir dans Le Volume même de L'Académie un de ses Membres appeler pour ainsi dire au public de son jugement et l'inscrive en quelque sorte en faux Contre des faits Constatés par six de ses membres nommé à cet effet."

33. RCL I/123–124 (April 1763): ". . . en retranchant tout ce qui peut y concerner M. Le Roy et leur dispute offrant à cet effet de la communiquer à tel commissaire que le Comité jugeroit à propos de nommer M. Duhamel a été chargé de cet examen."

34. "Réflexions sur quelques phénomènes cités en faveur des électricités en *plus et en moins*," MARS (1762), pp. 137–160 and 270–292.

35. Heilbron, *Electricity*, 362, and his "Nollet," 147.

❧ Chapter Seven ❧

Shaping Knowledge

*T*he Academy of Sciences was empowered to "examine and judge" the research submitted to it by its members and outsiders. The Comité de Librairie had the particular responsibility of vetting memoirs to appear in the Academy's *Histoire et Mémoires* and other of its publications. As we have seen, the Comité fulfilled its institutional responsibilities by seeing that papers met certain minimum standards, by having papers refereed by competent experts, by keeping track of the priority of work, and by adjudicating disputes when they arose. The cases examined in the last chapters suggest that the Comité and the Academy played an active role in shaping knowledge. Indeed, the Academy did not function simply as a dispassionate and august arbiter, but was an interested agent immersed in the making of science. The further examples of editorial intrusion investigated in this chapter show that the Academy in essence became an unattributed co-author of the papers it published.

The concerns of the Comité de Librairie with literary style and authorial tone provide an entrée into these issues. Because the author was perhaps not a native French speaker, the piece by one Radermacher received the comment, "Print, but correct the style." But the Comité said the same of Lafosse's article on a horse disease, Lafosse presumably a Frenchman.[1] Of a paper by Nollet, offered in the heat of his battle with Le Roy, the Comité insisted that he "soften several expressions that seemed too strong."[2]

The substantial editing to which some papers were subjected represents another dimension of the Comité's role in knowledge making. On a simple level, the Comité was concerned with the length of the *Mémoires*, and it regularly exhorted authors to shorten their papers. For example, it asked Macquer to abridge one Boucher's meteorological observations. Nollet took on the task of shortening Dutour's paper on diffraction.[3]

The Comité also intervened to alter the content of papers. Of Dutour's article on dissolving air in water, it twice ruled: "Print the first twenty articles; the rest is nothing."[4] Brouzet received permission to have his analyses of the mineral waters at Passy included in the *Savants Étrangers*, provided "the author stick to an analysis of the traditional waters and cut out what he says of the new ones."[5] Marcorelle's anatomical observations were approved, but the Comité called for "removing the two points mentioned in the report" and the paper appeared in the *Savants Étrangers* without the offending observations.[6]

In censoring material the Comité de Librairie imposed its own views on scientific subjects. Roëck's observation of a supposed satellite of Venus, for example, garnered a mention in the *Histoire* section, but the Comité advised the secretary "to take care not to suggest anything than might commit the Academy."[7] Evidence of this institutional intrusion is stronger regarding two papers on lightning by Cassini de Thury, Nollet, and other collaborators, about which the Comité twice ruled that it would "wait and see which side we are taking before making use of these articles."[8] The following examples provide more detailed insight into the role of the Comité as an active, sometimes partisan agent in the business of making knowledge.

Malouin on Air Temperature and Disease

The Comité's power to sway the course of knowledge can be seen in how it handled a series of medico-meteorological observations by Paul-Jacques Malouin (1701–1777), physician to the queen, professor of chemistry at the Jardin du Roi, professor of medicine at the Collège Royal, and an associate chemist at the Academy. Malouin published nine papers in the *Mémoires* from 1746 through 1754 that linked epidemic diseases to air temperature. In his initial reports Malouin made a strong connection between epidemic diseases and the ambient air and more particularly its temperature. He explained in a theoretical preface that began his report for 1747.

> In general our health depends more on the air than anything else. . . . When different qualities of the air are not proportional among themselves or not as they should be in each season, terrestrial beings for whom air is necessary are more or less affected, and these changes sometimes cause diseases, most often the ones common in a certain period that we call epidemic diseases.[9]

Malouin went on to elaborate a precise theory that had to do with air pressures inside and outside the body, digestion, the vacuum, the thinning of air on mountain tops, and like topics correlating the temperature and density of the air with specific epidemic diseases. Malouin thus placed himself in the camp of physicians who looked to environmental causes of disease rather than humors or temperaments possesed by individuals, but in emphasizing the role of air as an agency in disease, his environmental-ism was of a very specific sort. At the outset, the Academy seemed to back Malouin.[10] But not everyone agreed with him, for when Malouin's obser-vations for 1748 came before the Comité de Librairie, the Comité com-mented curtly: "Print, but cut out the theory of the air."[11] The change of heart may have had to do with the departure from the Comité de Librairie at the end of 1749 of Dortous de Mairan, a former secretary of the Academy and probably Malouin's supporter on the Comité.[12] Malouin modified his position slightly in his report for 1748, distinguishing epidemic diseases from ordinary illnesses "caused by some accident or by some particular vice or temperament," and he went on to add:

> In every country epidemic diseases have a common cause in those afflicted, and this cause can only be a bad quality or a corruption of one of the things used in common. Air and food are the most common things in the world and consequently are most often the cause of epidemic diseases. [I do not pretend] that these illnesses always depend on the temperature of the air. On the contrary, I am persuaded that certain epidemic mal-adies are caused by a hidden venom. . . .[13]

In 1750 and again in 1752 the Comité de Librairie continued to express its displeasure, twice asking Malouin to condense his reports.[14] While not departing from his commitment to the air as the principal cause of epidemics, in 1753 Malouin added other factors involved in the out-break of an epidemic: the quality of the local water, condition of fruits and grains, insects, and even "humors to be purged."[15] Malouin began his final report in 1754 by noting that his previous memoirs explained "how illness as well as life especially depend on air and nourishment," but he went on to enumerate a host of other factors affecting the salubrity of the Paris region where he made his observations: the number of animals; the disposal of human and animal sewage; the character of water supplies, including the Seine, public fountains, springs, and wells; the habits of Parisians in regard to drinking water and wine; matters of personal hygiene (especially femi-nine hygiene) and morals; annual rainfall, winds, barometric pressure, and, finally, the temperature and density of the air, the sole bases of epidemics in his first paper in 1746. Malouin might still cite Bacon and claim that "the vicissitudes of the air are the principal causes of the destruction of living beings," but he had so qualified his initial position of 1746 and 1747 that it no longer meant the same thing.[16] When it came time to summarize Malouin's work, Condorcet claimed that Malouin's theoretical interests encompassed "all the circumstances that impact human health: the air, its variations, temperature and humidity; local geography; food sources, the lifestyles of each region and of each order of society."[17] Doubtless other factors were at play, but the reluctance of the Comité de Librairie to accept Malouin's "theory of the air" changed the content and character of the public science that appeared in the *Mémoires*.

Controlling Access: The Difficult Career of Michel Adanson

In some meaningful senses, for knowledge to become knowledge it has to enter the public sphere. That the Academy of Sciences judged science and that the Comité de Librairie closely supervised the Academy's publications highlight the institution and its committee as bottlenecks through which claims had to pass on their way to becoming public knowledge. The control exercised by the Academy was not total or exclusive, to be sure, and even though the Academy repudiated the "science" of Mesmer, Marat, and other lesser lights, the views of these outsiders did make their way to the public, although not as sanctioned conceptions. From this perspective it might be said that the Academy shaped knowledge by defining a line between what was deemed genuine knowledge and what was branded as charlatanry. The more subtle point here concerns less dramatic instances where, by controlling access to an official press, the Academy and the Comité exercised institutional power to filter the unpublished, private science of academicians and so to mold a body of certified knowledge destined for specialist and general consumption.[18]

A notable case concerns the botanist, Michel Adanson (1727–1806), whom Charles Gillispie labels "a bold pioneer, a ferocious eccentric, . . . a maverick and an outsider . . . who won little hearing."[19] Adanson was hardly a model of civility or scientific collegiality, but why he was never fully accepted within the Academy is not entirely clear. Part of the explanation has to do with the fact that Adanson was not affiliated with the Jardin du Roi, that great center of botanical research and teaching presided over by Buffon and animated by the Jussieu dynasty of botanists. Part of the explanation has to do with Adanson's abrasive personality, his stubborn individualism, and a sense of grandiosity that kept him apart from the larger community of botanists. In any event, the story of Adanson and the Academy reveals efforts by the institution to mute or at least modulate the voice of one of its more contentious members.

Adanson studied at the *Jardin du Roi* in Paris and began his career as a naturalist with the *Compagnie des Indes* in Senegal from 1749 to 1754. The Academy named Adanson a correspondent in 1750, and in 1759 it elected him an adjunct in botany and gave him the honor of speaking at a public meeting. In his paper, "Plan of a General Work on Botany," Adanson "proposed that all methods and systems conceived for this science to date will fail."[20] His remarks must have been deemed inappropriate by the senior botanists of the Academy, for the minutes of the meeting make no mention of Adanson's presentation. When the Comité de Librairie reviewed the paper, it asked him "to change the tone of the work for publication," and, in fact, it never appeared.[21]

In 1763–1764 Adanson produced his major work, *Familles des plantes*, which again repudiated established taxonomy and nomenclatures and by implication attacked the botanical establishment at the Jardin du Roi. Possibly because of the cool reception he had previously received in the Academy, but possibly also because in the *Familles des plantes* Adanson argued in favor of the transmutation (read evolution) of the species, he published his major work outside the orbit of the Academy and without its imprimatur.[22] Through 1767 he authored only three minor reports requiring a review by the Comité de Librairie.[23] But friction developed again in 1767. The Comité de Librairie permitted his meteorological observations for 1766–1767 to appear "as an abstract in the *Histoire* section, but to suppress that which concerns mankind," suggesting that Adanson held controversial views about humankind. The paper did not appear until 1778, however, and then, not in the history section, as intended, but for other reasons in the *Mémoires*, with the most casual reference to humans.[24]

In March, 1767 Adanson read a paper, "On a Spontaneous Movement Observed in the Plant Called *Tremella*." The minutes for the meeting in question record Adanson's presence, but omit mention of his paper being read. The Comité de Librairie indicated that the work could appear as a memoir after "changing the preamble."[25] The sticking point concerned the absolute division of the plant and animal worlds, which the cautious academicians seemingly wished to defend. The opening paragraphs of Adanson's published paper are quite uncontroversial, and as reported in the *Histoire* section for 1767, "Mr. Adanson observes that one should not take the term 'spontaneous' rigorously and that the movement he has observed in the plant in question is still very far from the voluntary movements of animals."[26] In fact, however, Adanson writes in the body of the paper that "if any plant really participates in the animal and vegetable realms at the same time . . . without doubt it is the *Tremella*."[27] When the *Collection Académique* republished Adanson's paper on the *Tremella*, the editor's comment was more explicit and emphatic in drawing out Adanson's point: "It is commonly said that NATURE HAS BOUNDS AND LIMITS beyond which she does not stray in her productions, but ARE WE NOT TOO IMPATIENT SOMETIMES TO DEFINE THESE BOUNDS AND TO ASSIGN LIMITS?"[28]

The *Tremella* case formed part of a constellation of controversial scientific questions that emerged at mid-century. At issue were the separation of the plant and animal worlds, the fixity of species, and the nature of reproduction. Abraham Trembley's discovery of the fresh-water polyp in 1740 is the best known part of the story. The Swiss Trembley found that he could cut up the small hydra (*Chlorhydra viridissima*) into so many pieces and regenerate so many new organisms. He observed that the polyp reproduced asexually by budding. Experimentally he could fuse two creatures into one, graft a part from one onto another, or turn one completely inside out and have it metamorphose into a normal animal. These discoveries were quickly verified in the Academy of Sciences in Paris, notably by Réaumur, and they provoked amazement throughout the contemporary scientific world because they challenged received views of reproduction, the putative division between plants and animals, and notions of the fixity of the species. Regeneration dealt a near-fatal blow to preformationism, the widely discussed theory that explained reproduction by the successive development of one miniature generation stacked inside the other, and the whole issue raised the disquieting specter of materialism.[29]

The discovery of parthenogenesis in aphids, also in 1740, by Charles Bonnet compounded the uncertainty and developing sense of crisis caused by the hydra. Experiments concerning decapitated snails regenerating their heads constituted yet another facet of this unsettling research. The Academy was loath to venture far into these uncertain waters, and snail experiments in particular were not well received in the Comité de Librairie. Some of Lavoisier's early work involved regeneration experiments using snails, and in 1768, seemingly to avoid controversy, the Comité de Librairie voted to delay publication of Lavoisier's report on his snail experiments. In 1777 the Comité simply "suspended" Hérissant's paper on the subject.[30]

Adanson and the *Tremella* are to be situated in this context of controversy. As his published memoir of 1767 demonstrates, in the 1760s

Adanson leaned toward seeing the *Tremella* as intermediary link between plants and animals. The manuscript of Adanson's 1767 *Tremella* paper is missing, but an extant draft he began for a prize competition for 1768 sponsored by the Académie des Sciences, Belles-Lettres et Arts of Rouen allows us to pinpoint Adanson's views at the time "on whether noticeable and distinct limits separate the three realms of animal, vegetable, and mineral, or whether an unbroken chain links them into a larger whole." In his draft response Adanson wrote that "intermediate beings make up the series or the immense chain," and in a letter of 1767 he stated that "the distance is not great between certain animals such as the polyp and certain plants."[31] Adanson's views in the 1760s require this amplification because by 1775 Adanson had changed his opinion about the *Tremella*. Prompted by new work by Italian investigators, Adanson returned to experiments on the *Tremella*, this time classifying the species unequivocally as a plant.[32]

Adanson's change of heart regarding the absolute separation of plants and animals probably had to do with his mounting and bitter opposition to Linnaeus and Linnean taxonomy. Adanson likewise changed his earlier ideas about the transmutation of species and came to oppose Linnaeus on this point, too. But when the paper he read to the Academy in 1769, "New Experiments on the Question of Whether Plant Species Change," came before the Comité de Librairie, the group had serious qualms about it. The Comité ruled: "Print, after having had [the paper] read by M. de Jussieu who will remove all polemic from it," and the next month its registers repeat that "M. de Jussieu was asked to re-read M. Adanson's memoir and to eliminate everything from it that could become controversial."[33] The Academy publicized views under Adanson's name on the fixity of the species that differed from what Adanson believed in the 1760s, but without having access to a manuscript original, it is hard to say whether in 1772 Adanson subscribed to the credo attributed to him, or even whether it was his or Jussieu's:

> Mr. Adanson therefore believes himself authorized to reject absolutely the creation of new species. He remains in agreement that by means of artificial insemination—which in any case can only happen between individuals of the same species or closely related species—one can obtain variations and singular monstrosities, but not changes in species. These changes are hard to engender in plants; it is very difficult to bring them about in trees [sic] . . . But regardless of how large they are, these changes are not for all that changes in species . . . It seems

that the transmutation of species does not occur any more readily in plants than in animals. Even Nature's mistakes occur within certain bounds, beyond which everything reverts to the order pre-established by the wisdom of the Creator. It is with this judicious reflection that Mr. Adanson ends his memoir.[34]

Be that as it may, matters came to a head for Adanson and the Academy in 1775. Given his enlarged curiosity and plans, Adanson must have become exasperated with the pettiness of the responses he received because in 1774 he requested that the Academy examine "his works on all aspects of natural history relative to this science, to philosophy, and to metaphysics."[35] Adanson had been working for more than two decades on a universal encyclopedia of botanical knowledge, and in part his request for a formal review had to do with verifying that he had extensive files on the project.[36] In part, too, Adanson was asking the Academy to legitimate him personally and to ratify his work generally. The Academy obliged the request with a five-man investigating committee. Its report of March 1775 was damning:

> Sad, but true: All the aid that one might anticipate from Sovereigns who protect the Sciences could not produce for the public all the riches promised in his plan. From this observation we have concluded that Mr. Adanson should devote himself to his own contributions and make these public, so that in the future one can count him among those who will have contributed new observations to the mass of facts that will serve as the basis and material for this great work.[37]

Grandjean de Fouchy had the kindness to soften this rejection somewhat in providing Adanson a certified extract of the Academy's judgment:

> It would be desirable if, after publishing his prospectus, Mr. Adanson would finally start to make public the precious discoveries made in his travels that remain buried in his files. He should release the illustrations and descriptions of the animal specimens in his possession. For the publication of these works of great use to natural history he merits the protection of government.[38]

The Academy had had difficulties with Adanson all along, and all along it carefully monitored whatever he tried to publish through its offices. In March 1775 the Academy repudiated Adanson completely.

Later in 1775 the Comité de Librairie rejected his paper on monstrous pigeon eggs; in 1777 another piece of his earned an implacable "wait."[39] Adanson enjoyed a public reprise of sorts in the 1780s when the publication schedule of the *Collection Académique* caught up with events in the Academy from the decade before. But the 1775 encounter with the controlling power of the Academy seems to have taken the wind out of his sails.[40] Although he remained a pensionnaire botanist in the Academy until its closure in 1793 and a member of the Institut de France from 1795 until his death in 1806, Adanson had only one more paper published in the volumes of the pre-revolutionary Academy: the second part of an innocuous report in 1778 on the white gum tree from Senegal based on his trip to Africa nearly three decades earlier.[41] The man had been tamed by age and by the Academy's power to "examine and judge." Adanson published a total of only seven papers in the *Histoire et Mémoires* of the Academy in twenty years between 1759 and 1778.[42] The scant knowledge made public in the Academy's published pages is poor testimony to what Adanson had to say, but it speaks volumes about the Academy and the Comité as winnowing traffic along the road to knowledge.

The Comité Overruled

The difficulties experienced by Charles-Marie de La Condamine in getting his views on inoculation before the public illustrate the limits of the Comité de Librairie to control the production and dissemination of science. In this instance, the Comité pushed its powers too far and was overruled by the full Academy.

The technique of inoculation against smallpox was introduced into Europe from Turkey in the 1720s. Inoculation (or variolation) involved inducing a case of smallpox. (Inoculation is not to be confused with the much safer vaccination with cowpox that confers an immunity against smallpox, a discovery made in 1796 by the English physician, Edward Jenner.) Patients inoculated with smallpox usually experienced mild cases compared to those who contracted the disease naturally, but the procedure remained a risky and controversial one through the 1750s and 1760s in France. The duc d'Orléans had his children inoculated in 1756, and the successful outcome made inoculation fashionable in certain circles. But the Paris medical faculty remained divided over the safety and advisability of inoculation, and only after Louis XV died of smallpox in 1774 did the question resolve itself in favor of inoculation. In the meantime a bitter struggle unfolded.

Charles-Marie de La Condamine was not a medical doctor, which may have been the root cause of his difficulties, but he nonetheless became a leading proponent of inoculation. La Condamine read two papers on inoculation against smallpox at the Academy in 1754 and 1758 and published them in the *Mémoires* without apparent difficulty, perhaps because of the publicity surrounding the duc d'Orléans' children. A third paper, reporting on developments between 1758 and 1764, ran into trouble, however. It was scheduled for presentation at a public meeting of the Academy in 1764, but La Condamine was forestalled, apparently for want of time. He read the paper in closed session in late November 1764, and when the Comité de Librairie reviewed the paper, it ruled that La Condamine's work "will not be published either separately or in the *Mémoires* except after the report of Messrs. Camus and Petit."[43] About a paper by one Gatti on "prejudices against inoculation," the Comité expressed itself later that same month in an even nastier tone: "Not to be printed in any fashion." It rejected another inoculation paper by La Condamine the following summer with the terse: "Nt." [Néant/nothing].[44] The Comité de Librairie had become the locus of anti-inoculation forces within the Academy.

Three years elapsed, and Camus and Petit did not report. Finally, La Condamine succeeded in getting his protest brought to the Comité:

> Upon which, the Comité having deliberated, it was decided that . . . it is impossible to place this memoir in one or the other of these volumes [for 1764 or 1765], and anyway [because] this work by Mr. de La Condamine contains only excerpts from different published works on inoculation, it would seem more fitting to insert it in the public press than in the *Mémoires* of the Academy.[45]

La Condamine took his fight to the full Academy, where not without difficulty, he succeeded in getting Camus and Petit's report read on January 9, 1768. The report was overwhelmingly positive, concluding:

> We believe that this memoir by Mr. de La Condamine not only contains nothing that could hinder its publication, but on the contrary that it contains a large number of useful points which bring honor to the author and which make it worthy of a place in the collection of memoirs published by this Academy.[46]

Rather than acceding to the judgment of the referees, however, the Academy put off deciding whether to publish La Condamine's memoir, but at the same meeting it changed the membership of the Comité de Librairie. The abbé Nollet and S.-F Morand replaced L.-C. Bourdelin and d'Alembert as annual members, with d'Alembert continuing on the Comité as sub-director for 1768, and a Jussieu named the replacement permanent member for Duhamel de Monceau, who moved up to director for that year. Still, neither the Academy nor the Comité de Librairie took any action. Two weeks later, January 20, 1768, La Condamine appealed once more to the full Academy:

> My paper was twice read in our private meetings without anyone opposing its publication. Yet, the Comité de Librairie decided that it would not be published without a positive report from Messrs. Camus and Petit . . . When I asked the members of the Comité for a decision, I was told only that the report was not done . . . I could have published this memoir [elsewhere]. I was strongly urged to do so from several quarters, but I did not publish it either in its entirety outside the Academy or by abstract in the popular press. I waited for the publication of the *Mémoires* of the Academy . . . Mr. Camus brought the report to the Academy on the ninth of this month. At first the Academy refused to hear it, I pass over by what right. (This would have been the first time an academician charged to deliver a report did not have the freedom to give an account of it to the Company.) After I protested, it was read. More favorable than I had dared hoped, this report concluded in favor of publishing the memoir. . . .[47]

Finally, at the meeting that followed on Saturday, January 23, 1768, the full Academy, "having deliberated in the usual fashion upon Mr. de La Condamine's propositions made at the last meeting, judged . . . that with regard to the circumstances surrounding his memoir and the favorable report made by the reviewers, it would be published in [the volume for] 1765."[48] The Comité de Librairie then capitulated to the will of the Academy.[49] The Academy overruled its own Comité de Librairie and the members who opposed inoculation, and La Condamine's paper appeared in the 1765 volume without any indication of the difficulties surrounding its publication.[50] In the end La Condamine prevailed, not only against his scientific adversaries, but against their institutionalized control over the Comité de Librairie.

A final, small example reinforces what it meant for the Academy to control access to the scientific press. In April 1776 the Academy's director, J.-R. Tenon, presented on the part of one Jean Fortin, a celestial atlas for publication with the "approbation" of the Academy.[51] P.-C. Le Monnier and Messier reviewed it, and they reported that "the work merits the approval of the Academy and is worthy of being published under its privilege." Immediately on hearing the report, however, the full Academy "decided that the atlas in question could not be printed with the approbation of the Academy nor under its privilege and that the secretary would not issue any copy or extract of this report."[52] As it turned out, Fortin wanted to publish a new French edition of Flamsteed's stellar tables, but he had neglected to include anything in the projected work about the abbé Lacaille's observations of stars in the southern hemisphere. The omission was noticed, and for understandable reasons of academic and national pride (Lacaille had been sent to the Cape of Good Hope in the 1750s by the Academy) as well as the demands of more up-to-date astronomy, the Academy insisted that Fortin insert Lacaille's planisphere of southern stars in his work. The difficulty was easily remedied, and on Wednesday, June 5, 1776, Mr. Fortin entered the meeting room to show the Academy the engraving it demanded, "after which the Academy judged that the work . . . could be printed with its approbation and under its privilege and the secretary be permitted as a result to issue the necessary certificate to Mr. Fortin."[53] Later that autumn, the minutes recorded: "The edition of Flamsteed's atlas presented by Mr. Fortin."[54] The episode ended happily, but the Academy got the changes it wanted.

Controlling Dangerous Knowledge: The *Affaire* Paulet and the Cultural Politics of a Potato

One might object that the cases examined thus far were somehow exceptional or, conversely, that they represent the normal processes by which the grain of scientific truth is, so to say, winnowed from the chaff of error. But, what is one to make of the many papers by academicians simply marked, "Suspended"? What of Faujas' paper on volcanoes, Milly's on invisible perspiration, even Lavoisier's procedure for making oil of vitriol, all suspended?[55] What about the chemist and apothecary Sage's paper on sulphurous lead, which earned the blunt verdict "to suspend and demand explanations"?[56] The answers remain obscure, but the censorship power of the Comité could not be more evident. In two further instances, however, rich evidence reveals raw, political considerations lurking behind the decisions of the Comité de Librairie.

The Paulet affair, for example, began innocently enough in 1775 when the eponymous author read to the Academy a paper on a species of mushroom. While the reviewers considered the paper, Jean-Jacques Paulet made the mistake of publishing his paper in Rozier's journal, the *Observations sur la physique*, where it appeared as a full-fledged, twenty-two page memoir with two plates of engravings entitled, "On the Effects of a Mushroom . . . by Mr. Paulet, M.D., read at the Academy of Science."[57] This type of unauthorized publication with the misleading institutional association had occurred before, and the Academy drafted a formal complaint to the Lieutenant Général de Police, the official responsible for overseeing the book trade:

> For a long time now the Academy has had cause to complain that different public papers give notice of works that the Academy has definitely not published and which it would refuse publication without appropriate safeguards . . . A memoir, very interesting in itself, read and heard with pleasure by the Academy, has just been published as having been read at the Academy, which is true, but the memoir was still in the hands of the commissioners charged to examine it, and the Academy had rendered no judgment . . . Often, authors suppress the restrictions that the Academy has attached to its approbation, and works appear as approved in general by the Academy when they are only in part . . . Often, authors misuse the Academy's testimonials or anticipate its judgment to insert in the public press as having been presented to the Academy pieces that it has not approved or that it has approved with restrictions that authors ignore . . . The only remedy to this abuse is to enjoin the censors and those charged with monitoring journals and newspapers not to identify a paper as read or presented to the Academy without a certificate signed by its secretary testifying to the consent of the Academy.[58]

The Academy drafted a notice to journalists requesting them to insert a declaration that Paulet's paper had not been approved by the Academy, and Rozier duly complied.[59] Even in complaining, the Academy thought well of Paulet, and with the academician Le Roy as intermediary, the Academy let Paulet know that "it chose not to publish your paper, the value of which it recognizes, only for fear of spreading knowledge about a new poison with no known antidote."[60] Paulet's memoir, as it appeared in Rozier's journal, did in fact treat "the most dangerous mushroom in existence," and Paulet reported detailed experiments on poisoning dogs that he and A.-A. Parmentier conducted that effectively provided instructions on how to poison someone with *Fungus phalloïdes annullatus*. So, the Academy's stance was perhaps not just a sense of wounded institutional pride in seeing its name connected with something it had not formally approved. The case suggests an additional responsibility and covert role for the Academy as a guardian of knowledge and public safety on behalf of the state.

The Academy invited Paulet to re-read his paper so that a report could be delivered, and over the course of three meetings in 1776 Paulet presented an expanded treatment of several species of mushroom.[61] The Academy appointed L.-G. Le Monnier and A.-L. de Jussieu examiners, and their substantial report followed a month later. The reporters paid particular attention to the poisonous qualities of the mushrooms in Paulet's taxonomy. They noted, for example, that "Mr. Paulet describes seven other species of *Volva*, among which there are some very pernicious ones and which were the subject of the memoir he read previously to the Academy . . . The Princess de Conti and several other people were once victims of this kind of mushroom poisoning." With the proper procedures now observed, the examiners concluded that Paulet's paper "merits the approbation of the Academy and should be published in the *Savants Étrangers* series."[62] The Comité de Librairie rejected Paulet's paper, however, nominally because of its length and the number of its engravings, although one suspects that it did so rather as a guardian of public safety.[63] Citizen Paulet's *Traité des Champignons* appeared sixteen years later in 1793, after the Revolution had lifted press restrictions and after the Comité de Librairie had ceased to exist.[64] In this case the Comité de Librairie acted forcefully and successfully to suppress knowledge.

A final remarkable instance allows us to see the behind-the-scenes processes that eliminated work from the public record because of a threat to public safety and morals. To wit, on March 26, 1763, the Comité considered a paper, "Remarks by Mr. De la Ruë on the potato with two roots growing on Île Bourbon" in the Indian Ocean. The Comité voted to "suppress [the paper], given the danger of publication."[65] What could have been so dangerous about this "potato"? Fortunately for us, the letter from De la Ruë, copied into the Academy's minutes, suggests the answers.

The letter shows De la Ruë translating what he has to say across the cultural divide separating the medical establishment in Paris from botanico-medico knowledge systems of communities in Africa and the southern Indian Ocean. It is as if De la Ruë learned about the potato with two roots in a colonial patois and then, almost as a medical anthropologist, he set that knowledge over into a language and a set of conceptual categories his fellow European physicians could understand. This translation and the rhetoric of legitimatization it involved were all the more delicate because the potato with two roots embodied knowledge that was taboo in France.

After invoking Providence to explain the benevolent distribution of plants across the globe, De la Rüe attempted a scientific, botanical description of the plant he knew only as the potato with two roots. (It may have been a variety of mandragora or mandrake.) De la Rüe says that he learned of the plant through the natives ("les gens du pays") who used the plant to drain tumors. The medicalized context of the report is noteworthy. De la Rüe evoked the horrors of contemporary obstetric surgery, remarking that when women in France retain a dead child in the womb, "no means have been found until now other than the use of instruments invented by the medical arts for an operation so painful and accompanied with such deadly sequelae that decency requires I remain silent." He then continued: "A poultice made of this plant produces the same results as the operation, virtually without discomfort and without the fear of any untoward consequences."[66]

De la Rüe pursued the point with a revealing series of self-described "experiments" to demonstrate the effectiveness of the plant as a means to relieve the suffering of women. In the first one, a white woman ceased labor without giving birth. Having determined the child to be dead, the doctor applied the poultice and shortly thereafter "the patient delivered both a stillborn child and the placenta." In the next case, a black woman retained the placenta after the birth of her child: "A single application of the poultice of this same plant expelled the placenta after an hour and a half." In a third, "animal experiment," "I applied the same poultice to the abdomen of a goat two months pregnant. Three hours following the application, she aborted two kids which lived a half an hour." De la Rüe admitted that the plant could be misused "with a living foetus," but he insisted that "these findings should not remain private, given the advantages that the Public might take from them."[67]

This case exemplifies the flow of knowledge from the world of local practices on the colonial periphery back to the European center and incorporation into European knowledge systems. What were the learned men in the Academy to make of this communication and the knowledge it conveyed? Rüe was a genuine French doctor (or so it seems), and he wrote with the authority and professionalism of the same. That he transformed "native use" into Western science through a graded series of "experiments" must have made his claims seem even more solid and straightforward to the academicians in Paris. (That De la Rüe and the Academy would agree on the scientific categories and an experimental mentality reinforce the point.) That he introduced a therapeutic spin by evoking an alternative to the horrors of contemporary obstetrics would lead one to think that French science and medicine would have welcomed this botanical resource, making this an undramatic instance of the transmission of knowledge across cultural boundaries. In the event, matters turned out otherwise. After being read in the full Academy, De la Ruë's paper went to the Comité de Librairie where it was suppressed, as mentioned, not to see the light of day for over two centuries. What is more remarkable in this case: the colonial De la Ruë making knowledge through a cascade of experiments on a white woman, a black woman, and a nanny goat, or the eradication of this knowledge from the world of public science by the Comité de Librairie? The reason for rejecting the "potato with two roots" was never stated. It apparently had nothing to do with a disdain of native practices, a challenge to the authority of the rapporteur, or the legitimacy or truthfulness of his report. Rather, in this case the suppression of knowledge would seem to have sprang from a different quarter: clerical and state opposition to abortion.[68]

Notes

1. RCL I/136 (May 1764) and I/68 (November 1757): "Imprimé en Corrigeant le style."

2. RCL I/48 (November 1755): "qu'il addoucira plusieurs expressions qui ont parû trop Vives." About another piece by Nollet discussed at the same meeting, the Comité judged, "Imprimé et addoucy."

3. On Boucher's piece, see RCL I/38 (March 1754): "Imprimé et prier Mr. Macquer de L'abreger." On Dutour's, RCL I/90 (January 1760): "M. L. Nollet s'est chargé de L'abreger."

4. RCL I/16 (May 1751): "Imprimé les 20 premieres articles et le rest Nt;" RCL I/1 (February 1749). In point of fact, however, all fifty-seven articles of Dutour's paper appear in SE, vol. 2 (1755), 477–500.

5. RCL I/18 (August 1751): "Imprimé si l'auteur veux s'en tenir à Lanalyse des anciennes eaux et retrancher ce qu'il dit des nouvelles." In this case again the Comité's qualifications got lost in the shuffle, as Brouzet's paper appears in SE, vol. 2 (1755), 337–349 as "Analyse des anciennes eaux minérales de Passy, Et leur comparaison avec les nouvelles."

6. RCL I/165 (May 1767): "Imprimé en retranchant les 2 mentionnées au rapport." The two observations concerned people growing new sets of black, pointy teeth and a bull who passed two snakes via his anus; the latter observation was

rejected ". . . parce qu'elle ne paroît avoir été faite que par des gens de campagne peu instruit et très capables de se tromper ou de s'en laisser imposer;" PV 86 (1767), 92, 212–16; Marcorelle's paper, "Diverses observations anatomiques," appears in SE, vol. 6 (1774), 602–612.

7. RCL I/135 (April 1764): "histoire et prendre garde de rien avancer qui puisse engager L'acad.e." In fact, the HARS (1764) contains no mention of this paper.

8. RCL I/25 (June 1752): "Attendre qu'on ai vû quel party on prendra pour en faire usage." RCL I/25 (July, 1752): "attendre comme dessus."

9. MARS (1747), 563, 569.

10. See HARS (1746), 22.

11. RCL I/1 (February 1749): "Imprimé en retranchant La Théorie de l'Air."

12. Malouin was careful to cite Mairan in support of his conjectures in his first paper, MARS (1746), 151.

13. MARS (1748), 531. He elaborated the latter point in his somewhat incoherent report for 1751, MARS, (1751), 137–138, saying: "This venom in the air is usually different in the different years it appears. It is not the same from year to year, and consequently the illnesses it engenders are different, such that it is impossible to determine precisely the nature of their causes, no matter how attentive the most experienced doctors and *physiciens* be. . . . The secret cause of [epidemic] diseases sometimes arises from the soil."

14. RCL I/11 (August 1750); I/21 (January 1752).

15. MARS (1753), 35, 38.

16. MARS (1754), 495–498.

17. "Éloge de M. Malouin," HARS (1778), 57–65; here, 62.

18. On these points see above, Chapter 1 and Moran, *Silencing Scientists*, esp. 3, 33, 61.

19. *Science and Polity*, 153. On Adanson, see the two-volume collection edited by George Lawrence, *Adanson* that includes the substantial biography by Jean-Paul Nicolas, "Adanson, the Man," vol. 1, 1–121.

20. See HARS (1759), 115; see also PV 78bis (1759), 758; *Collection Académique, Partie Française*, vol. 12 (1786), 286.

21. RCL I/88 (November 1759): "Le prier de changer le ton de cet ouvrage pour L'Impression." The report on Adanson's talk written up in the Academy's *Histoire* for 1759 noted discreetly, "We could speak at greater length about this plan, but, because it should appear at the beginning of his treatise and M. Adanson having thought it superfluous to insert it in the *Mémoires*, we postpone a more detailed exposition until we give an account of the work as a whole;" HARS (1759), 115.

22. Laurent, "Classification," 461, notes Adanson's views on transmutation at the time of the *Familles des plantes*.

23. A positive notice of Adanson's book does appear in the *Histoire* section for 1763; HARS, 1763, 53–58. See also remarks in the *Collection Académique, Partie Française*, vol. 13 (1786), 260–262.

24. RCL I/161 (January 1767): "Par extrait dans Lhistoire et supprimée ce qui regarde les hommes." The *Histoire* for 1778 remarked that the paper "should have been inserted" in the volume for 1767, and a footnote explained that it had been "omitted" from the 1767 volume. (In fact, the Comité included Adanson's meteorological data to provide a basis of comparison with the record-breaking cold of 1776.) See MARS (1778), 425–32; HARS (1778), 1; PV 86 (1767), 19v.

25. RCL I/163 (March 1767): "Imprimé en changeant Le Préambule." See also PV 86 (1767), 67v.

26. HARS (1767), 75–78, here 78.

27. "Mémoire sur un mouvement particulier découvert dans une plante appelée *Tremella*," MARS (1767), 564–571, here, 571. See also *Collection Académique, Partie Française*, vol. 14 (1787), 181–183, where the editor comments, "Le hasard a offert à M. Adanson un mouvement spontané, & pour ainsi dire, animal, dans une plante. . . ."

28. *Collection Académique, Partie Française*, vol. 13 (1786) , 225: "On dit communément que la NATURE A DES BORNES & DES LIMITES, desquelles elle ne s'écarte pas dans ses productions; mais NE SE PRESSE-T-ON PAS TROP QUELQUE FOIS DE POSER CES BORNES & ASSIGNER CES LIMITES." Original emphasis.

29. On these points, see Dawson, *Nature's Enigma* and Baker, "Trembley." The ur-source is Vartanian, "Trembley's Polyp." On the larger background, see Roger, *The Life Sciences*, Part II and esp. 316–318.

30. On Lavoisier's observations, see RCL I/176 (July 1768): "attendre la suite des expériences." On Hérissant, see RCL II/66 (March 1777), where his paper is marked: "Suspendu." For more into this topic, see Smeaton, *"L'Avant-Coureur,"* esp. 223–225. See also HBI, AD 166, ALS Bonnet to Adanson, dated "À Genthod, près de Geneve, le [blank] Septembre 1769."

31. HBI, AD 277: ". . . les etres intermédiaires qui en comportent la serie ou la chaine immense." HBI, AD 206: ". . . il nià pas bien loin de certains animaux tels que le Polype a certains plantes." For background to this idea that that living (and some nonliving) things formed a single continuum or "Great Chain of Being," see Lovejoy's classic, *The Great Chain of Being*.

32. See HBI, AD 290: "Expériences sur la *Tremella* . . . [1775–1776]" and HBI, AD 291 "Nouvelles expériences sur . . . la *Tremella* [1775–1776]." See also Adanson's response to Bonaventura Corti of February 1775, where he writes, "mais je trouve malgré cela une gde. separation entre les 2 reignes;" HBI, AD 183.

33. RCL I/188 (November 1769): "Imprimé après avoir été Lu par M. De Jussieu qui en ôtera tout le polémique." RCL II/2 (February 1770): "M. De Jussieu a

été prié de relire Le mémoir Lu par M. Adanson le 15 novembre dernier et d'en ôter tous ce qui pouvoit devenir polémique." The Comité de Librairie met on February 2 and March 3, 1770 to consider papers on its books for November 1769 and February 1770. The de Jussieu in question was probably Bernard de Jussieu.

34. HARS (1769), 71–77, here 76–77. The paper in question, "Examen de la question si les espèces changent parmi les plantes: nouvelles expériences tentées à ce sujet," appears in MARS (1769), 31–48. The volume for 1769 appeared in 1772.

35. PV 93 (1774), 317: "M. Adanson a demandé des Commissaires pour éxaminer l'Etat actuel de ses Travaux sur tous les objets d'histoire naturelle relative-ment à cette Science, à la Philosophie, et à la métaphysique."

36. See HBI, AD 8, *Plan & Tableau de mes ouvrages*, an extract from Rozier's *Observations sur la physique 5* (1775): 257–276, which outline Adanson's phantas-magorical project. He claims that "friends" pushed him to ask for examiners from the Academy.

37. PV 94 (1775), 54v–59r: ". . . tous les Sécours qu'on pourrait attendre des Souveraines qui protégent les Sciences ne pourraient mettre le public en posses-sion de toutes ces richesses reunies sous un plan. De cette réflexion affligéante et vraie nous avons conclu que M. Adanson devait surtout s'occuper à détacher ce qui lui appartenait dans ces matériaux et a le rendre public afin que dans l'exécution d'un aussi vaste projet qui pourrait avoit lieu par la suite on puisse le compte parmi ceux qui auront contribué à mettre de nouvelles observations dans la masse des faits qui serviront de base et de matériaux à ce grand travail." See original of report signed by Le Roy, Guettard, Desmarets and de Fouchy, PS, Saturday, March 4, 1775.

38. HBI, AD 360; reproduced in Lawrence, ed., *Adanson*, vol. 1, 72.

39. RCL II/48 (March 1775); PV 94 (1775), 72; and RCL II/70 (December 1777).

40. See further references to Adanson's work in the *Collection Académique, Partie Française*, vol. 12 (1786), 228; vol. 15 (1787), 2; vol. 16 (1787), 169.

41. MARS (1778), 20–36.

42. Counting the first and second parts of his paper on the gum tree as one memoir. A two-page report by Adanson on the latitude of Podor on the west coast of Africa did appear in SE 2 (1755) which he earlier submitted in his capacity as a cor-respondant of the Academy. See listing of Adanson's papers, Halleux et al., *Les pub-lications*, vol. 2, p. 39.

43. RCL I/139 (November 1764): "ne sera imprimé ny à part ny dans le Vol.e qu'après le rapport de MM. Camus et Petit." See also PV 83 (1764), 373ff, 402, 403v, 408v.

44. RCL I/146 (July 1765). On the Gatti piece, see RCL I/140 (December 1764): "Non Imprimé d'aucune façon."

45. RCL I/169 (December 1767): "sur quoy le Comité ayant délibéré il a été décidé que . . . il est impossible de placer ce mèmoire ny dans l'une ny dans L'autre de ces Volumes et que d'ailleurs Cet ouvrage de Mr. Dela Condamine qui ne Contient que des extraits de differents ouvrages publiés sur L'Inoculation paroissoit plus propre à estre inseré dans Les Journaux que dans Les Mémoires de L'Académie."

46. PV 87 (1768), 2–3: "d'après cela nous estimons que ce memoire de M. DelaCondamine non seulement ne contient rien qui puisse en empècher l'impression, mais qu'au contraire il renferme un grand nombre de choses utiles, propres à faire honneur à l'auteur, et qui rendent l'ouvrage digne de trouver place dans le recueil des memoires imprimés de cette academie."

47. PS, January 20, 1768: "il fut lu deux fois dans nos assemblées particulieres, et personne ne s'oposa a l'impression. Cependant il fut décidé au comité qu'il ne seroit imprimé qu'apres l'aprobation de M.rs camus et Petit qui furent nommés Commissaires pour l'examiner. . . . Jai prié Messieurs du comité d'insister sur une reponse précise à ma demande, je n'ai pû en obtenir d'autre sinon que le raport des commissaires n'étoit pas fait . . . J'aurois pu alors publier ce memoire on me le demandoit avec empressement de plusieurs endroits; je na l'ai fait imprimer ni en entier, ni apart, ni par extrait dans les journaux. J'ai attendu l'edition des memoires de l'academie . . . M. Camus l'a aporté a l'academie le samedi 9 de ce mois; on a d'abord refusé de l'écouter, j'ignore de quel droit, c'auroit été la premiere fois qu'un academicien chargé d'un raport n'auroit pas eu la liberté d'en rendre compte a la compagnie; aussi sur ma répresentation la lecture en a été faite. Ce raport plus favorable d'ailleurs que je n'auroit osé l'esperer conclud à l'impression du memoire." See also PV 87 (1768), 8.

48. PV 87 (1768), 15v: "L'academie ayant délibrée suivant la forme ordinaire sur les propositions faites par M. Delacondamine à la séance dernier, il a été décidé . . . qu'en égard aux circonstances dans lesquelles s'étaient trouvé son mémoire et au rapport favorable qu'en avait faits MM. les Commissaires il serait imprimé en 1765."

49. See the marginal note, RCL I/169 (December 1767): "Depuis en Vertu de la Délibération de L'acad.e etant au registre sous le 23 Janvier 1768 il a été décidé que le mém. de M. DelaCond.e mentioné en cet article seroit imprimé en 1765."

50. "Suite de l'histoire de l'inoculation de la petite vérole, depuis 1758 jusqu'en 1765," MARS (1765), 505–532, and see the marginal note and footnote on 505 which give a sanitized version of this account of La Condamine's paper.

51. PV 95 (1776), fol. 126, meeting for April 24, 1776.

52. PV 95 (1776), fols. 131–132: "nous avons pensé qu'il méritait les suffrages de l'académie et digne d'être imprimé sous son privilege. [¶] L'académie ayant délibéré sur la conclusion du Rapport, a décidé que l'Atlas en question ne serait pas imprimée avec l'approbation de l'académie ni sous son Privilége et que je ne délivrerais ni Copie, ni extrait de ce Rapport."

53. PV 95 (1776), fol. 164: ". . . après quoi a arreté que cet ouvrage dont le Rapport a été fait le 30 avril serait imprimé avec son approbation et sous son privilége et m'a permis en conséquence de délivrer le Certificat nécessaire à M. fortin."

54. PV 95 (1776), fol. 296, meeting for November 16, 1776: "Edition de l'atlas de flamsteed presenté par M. fortin." The book appeared as *Atlas céleste de Flamsteed, approuvé par l'Académie royale des sciences, et publié sous le privilège de cette compagnie* ("revu et augmenté par P.-C. Le Monnier, F. Pasumot et l'abbé N.-L. de La Caille") (Paris: Chez F.-G. Dechamps, 1776).

55. Re Faujas, see RCL II/66 (March 1777); PV 96bis (1777), 391, 534–41v; PS, Wednesday, July 2, 1777. Re Milly, see RCL II/70 (August 1777); PV 96bis (1777), 455. Re Lavoisier, see RCL II/70 (August 1777); PV 96bis (1777), 455v.

56. RCL II/48 (February 1775): "Suspendre et demander des explications." See also PS, February 1, 1775. This seemingly innocuous paper by Sage appeared eighteen years later in the *Mémoires* for 1789, published in 1793.

57. *Observations sur la physique, sur l'histoire naturelle et sur les arts*, V (1775), 477–98. This piece also appeared separately as an off-print with a colored engraving (Paris: Ruault, 1775).

58. Undated, unsigned 2 pp. ms letter in folder labeled "Affaire Paulet," AdS, DG 31: "Il y a déjà Longtemps que l'acade s'est trouvée dans le cas de se plaindre en voyant paroître avec son attache dans les differents Journaux des ouvrages qu'elle n'avoit certainement pas publiés ou dont elle n'auroit permis La Publication qu'avec Les precautions nécessaires . . . un mémoire très curieux en luy même lu effectivement à L'Acade et qui y avoit été entendu avec plaisir vient d'estre publié comme Lu à L'acade ce qui est vrai mais il etoit encore entre les mains des Com.res chargés d'en faire L'examen. Lacade n'en avoit porté aucun Jugement. Plus souvant encour les auteurs ont supprimé les restrictions qu'elle avoit mises à son approbation et ont donné comme approuvé généralement ce qui ne L'etoit qu'en partie . . . souvent ils abusent du Certificat de L'acade ou préviennent son jugement pour faire inserer dans les journaux comme luy ayant été présentées des pièces qu'elle n'a pas approuvées ou qu'elle n'a approuvées qu'avec des restrictions qu'ils suppriment . . . Le seul moyen de remédier à cet abus seroit d'enjoindre à M.rs Les Censeurs et les Journalistes que sont chargés de L'examen des Journaux de n'y laisser insérer aucun mémoire comme L'eu ou présenté à L'acade ou approuvé par elle qu'on ne presentât avec La piéce L'approbation ou le Consentement de L'acade signé de son sécretaire."

59. See one page, undated draft note in folder labeled "Affaire Paulet," AdS, DG 31; and de Fouchy letter in Rozier's *Observations sur la physique, sur l'histoire naturelle et sur les arts*, VI (1775), 75, which concludes that "La compagnie n'a porté aucun jugement sur ce Mémoire."

60. See one page, undated note headed "A M Paulet" in folder labeled "Affaire Paulet," AdS, DG 31: ". . . elle ne s'étoit décidée à ne pas publier Vôtre

mémoire dont elle connoît d'ailleurs tout le prix que par la crainte de donner La Connoissance d'un nouveau poison Jusqu'à présent sans remède . . ."

61. PV 95 (1776), 126, 130, 133, meetings for April 24, April 27, and May 4, 1776.

62. PV 95 (1776), 165–172: "M. Paulet en décrit sept autres especes à Volva entier parmi lesquelles il y en a de très pernicieuses, et qui ont fait le sujet d'un mémoire qu'il a lu à l'academie . . . M. de La Princesse de Conti et plusieurs personnes en ont été les victimes . . . Nous croyons qu'il mérite l'approbation de l'académie et d'être inseré parmi ceux des savans étrangers qu'elle juge à propos de faire publier."

63. RCL, II/66 (March 1777).

64. Paris: Imprimerie nationale, 1793.

65. RCL I/122 (March 1763): "Remarques de M. De la Ruë sur la patate à deux rangs qui croît à L'Isle de bourbon; supprimé attendu le danger de La publication."

66. PV 82 (1763), pp 96–98: ". . . jusqu'à présent on n'a trouvé d'autre moyen que l'usage des instruments que l'art a inventés pour cette opération si douloureuse et accompagnée de suites si funestres que l'humanité m'engage à ne point ceder au public, l'effet d'une plante dont l'application seule en forme de Cataplasme produit le même effect que l'operation, presque sans douleurs, et sans qu'il y ait à craindre aucun évenement funestre."

67. PV 82 (1763), pp 96–98: ". . . la Malade fut délivrée et de l'enfant mort et du Placenta; . . . un seul Cataplasme de la même Plante l'a délivrée dans une heure et demi d'intervale; . . . trois heures après l'application elle a avorté de deux petites qui ont vecû une demi heure; . . . J'ai crû que ces expériences ne devoient pas être pour moy Seul, Vû l'avantage que le Public peut en tirer."

68. For a nuanced inquiry into contemporary views of gender and race raised by this example, see Schiebinger, *Nature's Body;* her "Exotic Abortifacients" provides a larger, comparative context for the present episode.

∾ Chapter Eight ∾

Taking Care of Business

*T*hus far, the Comité de Librairie has been depicted in its role as an august intellectual gatekeeper, but its less lofty functions as publisher and as an administrative arm of the Academy of Sciences were inextricably part of its operations. To dismiss these seemingly more mundane activities as somehow separate from the critical, scientific function of the Comité or as beneath attention is to miss an important dimension of what the Comité did as the publishing arm of the Academy.

The Comité de Librairie struggled with a range of practical problems in producing an annual volume of *Mémoires*. Controlling the size of volumes was a major factor that has been mentioned. When a tome threatened to become too large, the Comité occasionally divided an article into two parts to be published in different years.[1] At one point, the head of the *Imprimerie Royale* (the Royal Press) complained that the volume for 1764 had become too massive—over 700 pages not counting the *Histoire* section. The Comité reduced the number of memoirs, and "as a result," reported the secretary, Grandjean de Fouchy, "I went to the Press and withdrew [the papers], by dint of which reduction the volume will appear in a reasonable format of about 500 pages of *Mémoires* and 200 for the *Histoire*."[2] The same complaint arose for 1765 and the same expedient taken, which in turn only complicated matters for the volume for 1766![3] The number of papers for 1772 finally required publishing two tomes for that one year. Conversely, in the 1750s and early 1760s a dearth of papers meant that the Comité had to scramble to assemble a sufficient number to fill a volume. To bring the 1758 volume up to "its proper size," the Comité accepted Lalande's offer of "a long memoir on the latitude of stars" read in 1762 and 1763; Guettard's paper on the *Masilea* plant became destined "for 1761 if it is needed, otherwise print in 1762."[4]

Interactions with illustrators and engravers and concerns for the costs of engravings represent major preoccupations of the Comité de Librairie. Thirty-five percent of papers in the *Mémoires* were illustrated; for mathematics and botany articles, the figures stood even higher, at 60% and 58%.[5] Wood cuts very occasionally appeared on the page along with text, but 29% of all articles came with separate, engraved plates. These were expensive copper engravings that had serious implications for the Academy's budget.[6] The Comité monitored the expense of engravings closely, to the point where it rejected an anatomical paper by one Warnier for the *Savants Étrangers*, "even though worthy of the Academy," because of the number of figures it required. E.-J. Bertin's paper on the stomach of the horse provoked similar concerns because of the seven plates of engravings it required.[7] The Academy usually paid 30 *livres* for an engraving, and each took about two weeks to complete.[8] For more complicated figures the price rose to 40 or 50 livres, and for engravings of great complexity fees could range to 100 livres or higher. In one instance, the Comité approved the engraving of the mineralogical map of the Auvergne for 180 livres.[9] The number and expense of engravings became so great that in 1773 the Comité voted that "academicians will be required to communicate their figures in manuscript to see if it is possible to reduce their number."[10] The Academy and its printers split the costs of engravings, but nevertheless, the Academy's part of the bill for 1776 for drawing and engraving figures came to 20% of its total operating budget, exclusive of pensions.[11] At this point the Comité decided to shift the onus of paying for engravings entirely to its printers!

The Academy employed official illustrators (*dessinateurs*) and official engravers, who constituted key, if unheralded, personages in the Academy's production of science.[12] From 1749 to 1773 "le Sieur" Ingram was the Academy's official engraver. At one point Ingram requested "a pension of 500 livres to indemnify him for the low price paid for plates up until now and for the time he is often required to spend coming and going to the homes of the authors of various memoirs."[13] The king alone awarded "pensions," which were legal entitlements to an annual payment, so Ingram could not get one of these without more significant administrative action, but on its own authority the Comité saw no obstacle in granting Ingram a nonbinding "gratification" of 500 livres a year, a sum on par with the pensions received by junior academicians.[14] And because "the Royal Society of London requires its engraver to attend its assemblies," the Comité granted Ingram access along with correspondents to the Academy's otherwise closed meetings, so long as he attended in an official capacity.[15] Ingram likewise received page proofs so he could coordinate engravings with the explanatory text in the memoirs.[16] At one point he complained that authors gave him new figures to engrave after articles had been settled upon, which disrupted and delayed his work. The Comité therefore decreed that "academicians would be asked to submit all their illustrations with their memoirs and that no more would be accepted after printing had begun without a written order from the Academy."[17]

But Ingram was not the only person involved in producing illustrations. Occasionally, when the work load mounted, the Comité contracted out figures Ingram could not handle.[18] In 1773 the Academy took on one Fossier as its illustrator and Y.-M. Le Goüas as its engraver, replacing Ingram. The record discloses the presence of several women illustrators and engravers. Madeleine Basseporte, who was also affiliated with the court and with the Jardin du Roy, was the Academy's illustrator (*dessinatrice*) from 1742 at least through 1776.[19] She received 400 livres a year, and she, too, enjoyed the right to attend meetings. Other references mention the sisters, Elisabeth and Catherine Haussard, who did contract work on engravings.[20] A poignant letter of theirs asks the Academy for a raise: "We have engraved the plates for the volumes of the *Savants Étrangers* for many years, but that is becoming more and more difficult."[21]

Volumes had to be printed, and in seeing to the manufacture of books, the Academy and the Comité interacted with the artisanal worlds of eighteenth-century printers and the book trade (Fig. 7).[22] These deal-

Figure 7: An unsigned silver *jeton* of the guild of booksellers and printers. Obverse: Coat of arms of the corporation crowned; exergue with date of M.DCC.XXIII. Reverse: an open book beneath the sun with the legend, "EX UTROQUE LUX"—"Light from Each"; exergue: BIBLIOPOLÆ ET TYPOGRAPHI PARIS.IS"—"Booksellers and Printers of Paris." Feuardent #5264.

ings are seen primarily in the contracts the Comité negotiated with different printers. In 1751, for example, Anisson, then director of the Royal Press, bid to take over printing the memoirs, provided, among other conditions, "that the Academy appoint someone to read proofs in the absence of authors so that printing can proceed all year around and that authors accept that person's corrections."[23] In 1776 the Comité played off its long-time printer, Perron, against a new bidder, the famous Charles-Joseph Panckoucke.[24] Perron acquiesced to increasing the number of volumes he furnished gratis to the Academy to 130, but he balked at taking over the costs of engravings. Panckoucke, on the other hand, went along with the conditions, "as onerous as they are, because he already holds a large stock of the Academy's works."[25] The Comité relented somewhat, demanding only seventy-six free engravings for five years, and Panckoucke won the contract. The Academy made separate, but similar arrangements for the *Description des Arts et Métiers*, for the publication of papers that won its prize contests, and for other of its volumes.[26]

For their part, printers habitually complained about the size of volumes and delays in getting materials, notably engravings, from authors, the Comité, and le Sieur Ingram.[27] The problem of coordinating the editorial work of the Comité with the labor of the print shop became so acute in 1763 that a group of academicians met with Perron at his shop to hammer out new procedures. They concurred that in the future,

> After each meeting of the Comité, the secretary will inform the full Academy first and then each author of the memoirs to be printed. Authors will be required immediately thereafter to submit draft figures to the engraver who will work on them in the order they appear on a list kept to this effect by the secretary. The secretary will no longer accept any memoir for publication that is submitted to him without drafts of the illustrations. This arrangement seemed even more practicable, that by dint of these different submission dates, it will always be easy to see where delays originate and to make corrections. This was approved definitively in the Comité.[28]

The business difficulties faced by printers and booksellers are revealed in a further instance. In 1761 Messrs. Dessaint and Saillant were publishing the Academy's *Description des Arts et Métiers*. They reported to the Comité de Librairie that they had learned of a Dutch counterfeit edition of 500 copies. Their contract with the Academy permitted them to print 600 copies, 100 of which were destined for the

Academy itself, which left Dessaint and Saillant without "a means to stop this counterfeiting which is costing them dearly." They asked for permission to print an additional 500 copies "to be distributed in Holland at a modest price."[29] The Comité granted its permission. One wonders whether Dessaint and Saillant were telling the truth, or whether they were tricking the Comité de Librairie in order to double the number of volumes they could sell in Holland and elsewhere.

Last-minute changes also kept the Comité involved with the print shop. For example, the secretary reported that as a result of the late decision in 1761 to insert papers by J.-B. de Chabert and the abbé N.-L de Lacaille in the volume "currently in press, I brought the memoirs to the printer on the spot."[30] The Academy paid substantial sums for authors' corrections and these last-minute changes.[31] Then, having volumes bound represented a separate step in the production process, for which the Academy had to pay extra.[32]

The Comité de Librairie shared with the Academy responsibility for distributing complimentary copies of the Academy's publications, and at various points the Comité voted to send the *Histoire et Mémoires* and the *Connaissance des Temps* to sister societies in London, Saint Petersburg, Montpellier, Stockholm, and Haarlem. The *Académie française* and the *Journal des Savants* received these publications as well. Copies of the *Arts et Métiers* series went to institutions and to individuals who collaborated on particular projects, and in 1777 the Comité donated copies of the *Connaissance des Temps* to an expedition departing for China. As a rule, the Academy received only 100 free copies of its own publications from the printer, a number often insufficient to cover the required distribution to academicians and the court. As a result, to fulfill its obligations the Comité regularly had to go out on the market and buy additional copies from booksellers.[33]

In 1779 the Comité agreed to give the annual volume of the *Histoire et Mémoires* to the *Société Royale de Médecine* and to abbé Rozier.[34] The mention of the abbé François Rozier is worth noting. Rozier had launched his *Observations sur la physique, sur l'histoire naturelle et sur les arts* (also known as the *Journal de physique* or Rozier's journal) in 1772 as a monthly publication designed to overcome the notorious time lag in the appearance of learned society memoirs, not least those of the Academy, and to speed communications between and among active researchers. However, Rozier did not work in opposition to the Academy and its established procedures, as might be expected, but as a complement to them, and he was well received by the institution.[35] The relevant point here is that the Comité de Librairie used Rozier and his journal as another publication outlet for the

Academy. For instance, regarding a report on wasps found on the island of Guadeloupe, the Comité decided to describe the findings in the "*Histoire* and give the memoir to the abbé Rozier." Of another memoir, it ruled simply to "print it in Rozier."[36] These examples suggest an even closer working relationship between the Academy (through the Comité de Librairie) and Rozier than has been recognized to this point.

Then, annually or when the occasion arose, a delegation from the Academy traveled to Versailles to ceremonially present the latest volume of its *Histoire et Mémoires*, regally bound in leather and stamped with the royal coat of arms, to the king, the royal family, and other dignitaries at court. The minister in charge of the Academy would indicate when such visits were appropriate, and doubtless the Academy's president, vice-president, director, and subdirector led these visitations. Whether other members of the Comité de Librairie regularly participated is not known. The formal presentation to the king completed the production cycle of the Academy's publications. Typically, the deputation from the Academy then dined with the minister.[37]

Finally, in addition to its role as publisher, the registers of the Comité de Librairie reveal a group of people grappling with the many day-to-day issues faced by a major standing committee of the Academy. This executive function can be seen in the series of minor matters that came before the Comité: getting instruments in the Academy's collections repaired, allowing access to a deceased academician's papers, dealing with the academician who put a copy of his work in the Academy's library but did not give it to the secretary, or granting permission for people to pass through the Academy's apartments in the Louvre on weekends and holidays. The Comité consented to the latter request, provided that a Swiss guard be on duty and that its permission did not "extend to any servant except those necessary to serve or carry their masters."[38]

The Academy's customary presentation of its volumes to the king or this last image of servants carrying masters through the Louvre, like so much else encountered in this study, underscores that the Comité de Librairie, like the Academy of Sciences itself, was decidedly an institution of the Old Regime.

Notes

1. See, for example RCL I/7 (1750) and MARS (1746), 713–758 and (1747), 699–743 for two parts of a paper on the natural history of the Languedoc by the abbé Sauvages.

2. RCL I/159 (November 1766): "En Conséquence de la quelle Décision je me suis transporté à Imprimerie et . . . au moyen du quel retranchement le Volume se trouvera d'environ 500 pages de Mémoires qui avec les 200 d'histoire forment un format raisonnable." However reasonable, this step coincidentally created problems for the reverend A.-G. Pingré, whose paper on solar parallax was shifted to the 1765 volume of the *Mémoires*, thus making it look as if the English astronomer, Short, had priority over Pingré in an on-going dispute; see the delightful Pingré letter, "A Messieurs les Commissaires de la Librairie," dated February 27, 1767 in AdS, Dossiers Biographiques, Pingré. He concludes: "Sur ces considérations, Messieurs, qu'il me soit permis de vous conjurer de trouver dans votre esprit vaste, profond et Officieux quelque ressource qui puisse procurer dans la collection de 1764 une pauvre petite place à mon Mémoire sur la parallaxe du Soleil. . . ." His plea fell on deaf ears, but the Academy mounted a substantial defense of Pingré's priority the next year, HARS (1765), 77–82.

3. RCL I/169–170 (December 1767); I/159 (November 1766). The *Savants Étrangers* series faced similar problems; see RCL I/177 (July 1768).

4. Re Lalande, see RCL I/125 (June 1763): "ce qui mettroit le Volume à sa Juste Grosseur." Re Guettard, see RCL I/117 (September 1762): "Imprimé en 1761 si on en a besoin sinon en 1762." For other examples of this sort, see also RCL I/4 (June 1749); II/18 (January 1772).

5. McClellan, "A Statistical Overview," 32, Table 11.

6. On contemporary scientific illustration, see Knight, "The Growth of European Scientific Monograph Publishing," esp. 30–32.

7. Re Warnier, see RCL II/21 (May 1772): ". . . bien qu'il soit digne de l'académie." On the "affaire des planches" surrounding Bertin's paper, see RCL I/12 (December 1750), and MARS (1746), 23–54 with the accompanying engravings by Robert.

8. On these points, see RCL I/163–164 (March 1767) and PS, Pochette Générale—1776, "Mémoire des dessins que Mr. fossier dessinateur de L'academie a fait." A skilled tradesman, such as an engraver, typically earned 1–2 *livres tournois* a day in contemporary France.

9. For these cases, see RCL I/47 (November 1755); I/72 (March 1758); II/40 (April 1774); see also the fold-out engravings by Guillaume De la Haye accompanying Desmarest's article, MARS (1771), at 774. Various PS contain bills from engravers; for example, PS, Pochette Générale—1773, contains "Memoire des Planches gravées par le Gouaz dans le courant des années 1772 et 1773 pour le volume de memoires de l'Academie Royale des Sciences, de l'annee 1770," "Memoire des desseins qu'a fait Mr. fossier pour L'academie Royalle des Sciences qui servent aux memoires de 1773," and "Memoire des planches Gravées par la Gouaz dans le courant de lannée 1777, pour le volume des Memoires de L'Academie Royale des Sciences de l'année 1773."

10. RCL II/27 (January 1773): "Les académiciens seroient obligés de Communiquer leurs planches au Comité pour voir si L'on peut les réduire."

11. For 1776, the illustrator billed 1,664 livres; the engraver's bill came to 980 livres, of which the Academy paid 530 livres; and the Comité commissioned an additional twelve illustrations for 288 livres for a total expense of 2,482 livres. These figures are taken from PS, Pochette Générale—1776, "Memoire des dessins que Mr. fossier dessinateur de L'academie a fait" and "Mémoire des planches gravées par Le Gouaz dans le Courrant de lannée 1776. . . ." An overall budget figure of 12,080 livres is mentioned RCL II/67 (August 1774). See also, AdS, RCT, 16 (meeting for December 23, 1763), where in 1763 engravings alone constituted 15% of a budget of 12,500 livres; for 1764 the costs for engravings amounted to nearly 16%; for 1769 (RCT, 36), the figures approach 17% of the annual budget, exclusive of pensions. For comparative figures for the seventeenth century, see Stroup, *Royal Funding*, passim and Tables 5 and 7.

12. In 1707, for example, the Academy granted the request of M. de Chatillon, its "long-time" illustrator, to attend meetings in order to make sketches on the spot, provided that Chatillon also come to meetings when requested; see AdS, DG 31, document labeled "Règlements [1699 à 1738]," fol. 2v, under date for "Avril 13 1707." Claude Aubriet (1665–1742) succeeded de Chatillon. Duprat, "Les Dessinateurs d'Histoire Naturelle," provides a complementary perspective on the world of contemporary scientific engravings. See also the relevant entries in Bénézit, *Dictionnaire Critique*. The recent paper by Pinault-Sørensen, "Les dessinateurs de l'Académie," now provides the entrée into this topic.

13. RCL I/72–73 (March 1758): "Une pension de 500# qui puisse L'Indemnifer de la modicité du Prix auquel Les Planches ont été payées Jusqu'icy et du Temps qu'il est obligé d'employer souvent à aller et Venir chéz les auteurs des Differents mémoires." See RCL I/5 (November/December 1749) for Ingram's appointment as the engraver. Curiously, this Ingram is not listed in Bénézit, *Dictionnaire Critique*.

14. Mention of Ingram's "gratification" for 1761 shows up in RCT, 12 (meeting for February 19, 1763) and following years.

15. RCL I/72–73 (March 1758): "La Société Royale de Londres oblige son graveur à assister à ses assemblées. . . ."

16. RCL I/23 (April 1752).

17. RCL I/157 (September 1766): ". . . il a été décidé que Mrs. les académiciens seroient priés de remettre toutes leurs figures avec les mémoires et qu'on n'en recevoir plus aucune après L'Impression Commencé sans un Ordre par écrit par L'académie." See also RCL II/27 (January 1773) where this order is restated. At another point, because payments had been delayed, academicians Duhamel and Macquer offered to advance Ingram 200 livres against the later distribution of funds; see RCT, 43 (special meeting of February 13, 1771).

18. RCL I/153 (April 1766); I/154 (May 1766); see also RCT, 49 (special meeting for 17 February 1773).

19. On Madeleine [Magdeleine] Basseporte (1701–1780), see RCL I/72–73 (March 1758); II/67 (August 1774); RCT, 12 (meeting for February 19, 1763), where her pension is listed "par forme de Gratification;" she continued to receive this "gratification" at least through 1776, when the records of the Comité de Trésorerie leave off. On Madeleine Basseport, see also Poirier, *Femmes de science*, 348–349; Velut, *La Rose*, 42–43; Nicolas, "Adanson", 141; Duprat, "Dessinateurs," 457–460; Bénézit, *Dictionnaire Critique*; and AdS, DG 31, "Collection des Reglemens et Déliberations de L'Académie Royale des Sciences, par ordre de matière [1699–1753]," 34.

20. On the Haussard sisters, see RCL I/163–164 (March 1767); II/[72] (March 1778); their first names are taken from their engravings that appear in SE; see also RCT, 45 (Meeting of April 27, 1771). Only Catherine Hussart [sic] is listed in Bénézit, *Dictionnaire*.

21. Undated ALS in AdS, DG 31: "Imprimeurs de l'Académie royale des sciences": ". . . La Graveur des planches du Volume des Scavants Etrangers, que nous Executons depuis Plusieurs années, devient de plus en plus difficile. . ."

22. For a justifiably renowned portrait of this world, see Darnton, *The Great Cat Massacre*. See also Minard, "Agitation." For an entrée to the literature concerning the history and material culture of the book, see Darnton, *The Kiss of Lamourette*, esp. chapt. 7, "What Is the History of Books?"

23. RCL I/15 (March 1751): "Que L'académie Commette quelqu'un pour lire les épreuves en L'absense des autheurs afin que L'Impression aille toujours le même train dans tous les temps de L'année ou que les autheurs s'en rapportent aux corrections qu'il commettra offrant en ces cas s'il se trouve des fautes dans L'Imprimé qui ne soient pas dans La Copie de faire refaire les feuilles à ses depens." The Comité accepted these conditions.

24. On Panckoucke, see entry in Sgard, *Dictionnaire des Journalistes*, 295–296.

25. RCL II/63 (November 1776): "M. Pankouke Lui a laissé entendre qu'il accepterait les Conditions proposées par L'academie quelques onéreuses qu'elles fussent; par ce qu'il avait des Fonds Considerables en ouvrages de L'académie."

26. See RCL I/84 (June 1759); I/90 (February 1760); I/94–95 (May 1760). For other contracts, see AdS, DG 31: folder labeled "Imprimeurs de l'Académie royale des sciences," which includes, among others, a contract dated August 14, 1734 with "sieurs Martin, Coignard fils & Guerin l'ainé" for publishing the *Histoire et Mémoires*. PS, August 1727, contains an "engagement" dated "a Paris le 17e Aoust 1727" with the printer, Montalan, to print a variety of the Academy's volumes. For later negotiations with printers, see PV 99 (1780), fol. 9v.

27. See RCL I/8 (March 1750); I/75–76 (June 1758); I/155 (July 1766).

28. RCL I/131 (December 1763): "Il a été décidé qu'à L'avenir à la seance que [prévoit?] le Comité de chaque mois Le sécretaire Indiqueroit d'abord en pleine acade. et ensuite en particulier à chaque auteur Les Mémoires qui auroient été destinés à L'impression que les auteurs seroient tenus de remettre immédiatement après leurs figures au Graveur qui s'en chargeroit sur une feüille tenue à cet effet par le sécrétaire pour y travailler successivement et que le sécretaire ne recevoit desormais aucun Mémoire pour LImpression qu'on ne luy remis en même Temps L'épreuve des figures gravées qui le Concernent Arrangement qui a paru d'autant plus utile qu'au moyen des dates de ces Différentes remises il sera toujours aisé en cas de retardissement de voir de quelle part il vient et d'y remedier Ce qui a été arrêté diffinitivement au Comité."

29. RCL I/101 (January 1761): ". . . le seul moyen qu'ils eussent trouvé pour empêcher cette Contrefaçon qui leur feroit un tort Considérable étoit de faire passer eux même en holland 500 éxemplaires pour y estre distribuéz à un prix modique."

30. RCL I/103 (April 1761): ". . . ces deux écrits concernant Le passage de Vénus sur le soleil seroient inserés dans le Volume de 1756 actuellement sous presse. en Consequence de quoy je les ai portés à L'Imprimerie sur le champ."

31. See RCT, 17 (meeting for December 23, 1763), where one Béjor received 600 livres for author's changes. APS Fougeroux papers, Botanical Mss., 580/D881/no. 14 contains marked-up page proofs for two of Fougeroux's memoirs, with a note added by Fougeroux: "je prie Monsieur Vercavin de m'envoyer une seconde Epreuve de Ce memoire," indicating that papers sometimes went through two rounds of proofs.

32. RCT, 20 (meeting for January 12, 1765), where "La V[euv]e Pasdeloup" received 381 livres "pour relieures faites pour L'Acade." See also PS 1773, where the widow Le Monnier earned 227 livres for binding fifty-six volumes of the *Description des Arts*; those she bound in "marocq. armes et dentelles" cost the Academy 15 livres apiece.

33. On these examples, see RCL I/70 (February 1758); I/77 (August 1758); I/104 (April 1761); I/128 (August 1763); I/157 (September 1766); II/46 (November 1774); II/71 (December 1777). See also AdS, DG 31, which includes a twelve-page notebook, "Extraits du Registre des Comitès de Librairie, qui accordent un exemplaire des ouvrages de l'Academie" and an undated memo of S.-F. Morand ("Pour Le Comité de librairie") that notes the fifty-two extra copies of the Academy's publications to be purchased. The author remains ignorant of the cost of a typical volume, but at one point the Academy expressed concern over what it perceived as the high cost of its *Mémoires*; see PV 99 (1780), fol. 5. On a related matter, where academicians were offered off-prints rather than the Academy dumping volumes wholesale to bookstores, see RCL I/173 (March 1768).

34. RCL II/[75] (June 1779).

35. Rozier produced a substantial index of the Academy's publications in 1775–1776, and in 1783 he became a correspondent; see McClellan, "The Scientific Press in Transition." See also entry in Sgard, *Dictionnaire des Journalistes*, 327–328, and contrast Meadows, *Communication in Science*, 70.

36. For these examples, see RCL II/49 (March 1775); II/53 (April 1775): "histoire et donner le mém. à M.L. Rosier … L'imprime dans L. Rosier." See also RCL II/39 (March 1774); II/[75] (May 1779). By the same token, contrast PV 99 (1780), 8, where a paper is dropped from consideration because of its prior publication in Rozier.

37. See PV, passim. The author has not found any account of these expeditions from the Academy of Sciences, but an almost comic report of 1774 details a delegation from the Académie des Inscriptions et Belles-Lettres traipsing around Versailles looking for notables to whom it could offer its volumes; see Institut de France, Archives de l'Institut, Cote: A: 66, "Registre des Assemblées & Délibérations de l'Académie Royale des Inscriptions & Belles Lettres pour l'année 1774," 147–148; see also the same academy's "Registre" for 1771, p. 61 and 1778, p. 31.

38. On instruments: RCL I/141 (January 1765), I/177 (July 1768); on access: RCL I/14 (February 1751), I/93–94 (April 1760). On the library instance: RCL I/48 (November 1755). On permission to use the Academy's rooms: RCL II/66 (February 1777): "cette permission ne s'etendit à aucun domestique accepté Ceux qui seraient necessaires pour maintenir ou porter leurs maîtres."

᧤ Chapter Nine ᧢

Conclusions

*T*his study has explored a key episode in the making of modern science. In early modern Europe inquiry into the natural world encompassed a wonderful diversity of activities, approaches, institutions, and individuals: bookish Aristotelian professors in universities, court-sponsored intellectuals and patronage-dependent artists, scholars of independent means, military men, physicians, lawyers, priests, printers and booksellers, wandering and sedentary alchemists, and mystics. Some investigators were devoted to mathematics, some to experiment, some to mechanism, and some to mysticism of various shades. Some found truth in the wisdom of the ancients; others in the novelties of the moderns. Clearly, through the middle of the seventeenth century there was no single, agreed-upon enterprise that constituted inquiry into nature. Natural philosophy wore a thousand faces.

By the beginning of the nineteenth century, although still hardly a unitary endeavor, the enterprise of science had become more narrowly defined and a more clearly recognizable social and intellectual affair.[1] The modern scientific disciplines were taking shape, and each came to be represented by a corresponding scientific society devoted to a specialized branch of knowledge. Universities became endowed with laboratories and research missions, and for the first time professional organizations of scientists entered onto the historical stage. Indeed, the coining of the English word "scientist" in 1840 is usually taken to exemplify the solidi-fication of recognizably modern science out of its more diverse and flamboyant roots in early-modern history.

From this overarching perspective the years from 1650 through 1800 witnessed a gradual, but remarkable, transformation of the scientific enterprise from its early-modern to its more modern incarnations. This transformation played itself out in a myriad of specific settings, of course, and historiographically it would be a mistake to attribute unity or inevitability to the historical trajectory of science in the later seventeenth and eighteenth centuries. Still and all, the scientific enterprise had a vastly different social, institutional, and cognitive character at the end of this period than it did at the beginning. The great intellectual syntheses found in Newton's *Principia* (1687) and his *Opticks* (1704) initiated new eras of research in the physical sciences.[2] The institutionalization of science in learned societies like the Royal Society of London and the Académie Royale des Sciences in Paris introduced a major structural novelty. As further evidenced by the spread of astronomical observatories and botanical gardens, increased support by European governments likewise provided new underpinnings for contemporary science. The enhanced opportunities for individuals to pursue full-time careers in science was a noteworthy development. Not least in this connection, the appearance of scientific journals constituted another element in the gradual emergence of modern science over the course of the seventeenth and eighteenth centuries.

The present study has been built on the premise that it was within the learned societies of the seventeenth and eighteenth centuries that autonomous cadres of scientific experts first achieved formal, direct control over the published word of science. The Académie Royale des Sciences and its Comité de Librairie tell the tale in this one, groundbreaking instance, where an abundance of archival and other resources has allowed us to follow the development of specialist control over science in fine detail. We have seen the Comité de Librairie apply strict standards to works approved for publication. We have seen the Comité institute peer review of work and demand revisions based on the judgment of its reviewers. We have seen the Comité act as the guardian of priority and judge in cases of priority disputes. We have seen the Comité as an active, albeit behind-the-scenes agent in the production of knowledge. And, we have seen the Comité as a beleaguered publisher coping with the concrete problems of turning out volumes of scientific memoirs over the course of a century.

The "why" of it all—why the producers of science became the gatekeepers of science—is not difficult or time consuming to explain. The Academy and the Comité were chartered by royal authority and functioned as part of a larger bureaucracy devoted to the maintenance and expansion of the French state and the Bourbon monarchy. The Academy provided the State with useful technical expertise in a number of domains, and in return the state subventioned the Academy (and its publications) and allowed it to become self-governing, at least in matters of science.[3]

This study has documented how the academicians of the eighteenth-century Academy exercised their authority in the arena of scientific publishing.

Examining the Comité de Librairie illuminates a heretofore little-known, but significant aspect of the history of the Academy of Sciences in the eighteenth century: the day-to-day operations of a principal executive committee that had a hand in shaping the history and reputation of the Academy. As already noted, the Paris Academy of Sciences was the single most important institution treating the natural sciences in the eighteenth century, and a rich historiography now embellishes our knowledge and understanding of the organization. A study of the Comité de Librairie may represent a contribution to that historiography, but only if we are explicit about how it clarifies the general history of the Academy. Here again, the conclusions are fairly straightforward. They speak to the increasing autonomy and professionalization of science that is otherwise evident in the general history of the Academy in the period. Moreover, what we have seen reinforces the thesis put forward by Roger Hahn in his study of the Academy: that, as the institution charged to certify truth, eliminate falsehood, and quarantine charlatanism, the Academy became increasingly identified as a bastion of elitist privilege as the Revolution approached.[4] The notorious cases concerning Mesmer and Marat did not come before the Comité itself, but the hundreds of rejections it handed out no doubt occasioned resentments that added to the reputation of the Academy as a severe and undemocratic arbiter. In the Nollet–Le Roy dispute seen above, the Comité's refusal to "let the public decide" provides sharp evidence of its mentality in these regards. But a further point has to be underscored here. That is, our sense of the Academy is enriched when we realize that the Comité de Librairie served as the mechanism to sort out *internally* the productions of academicians. Some works received the literal and metaphorical imprimatur of the Academy, others did not. Just because someone was an academician did not mean that their work was *ipso facto* worthy of being made public, and in the more dramatic instances we have examined, no less than the many mundane cases, the Comité de Librairie was the locus where the research of academicians received scrutiny by their peers. The setting was private and discreet, and institutional diplomacy was called for, but the judgments of the Comité did not lack rigor on that account. The institutional commandment to "examine and judge" demanded no less.

Along these lines, this study has shown that the Comité de Librairie imposed certain norms on the works it certified and published. Those norms included that papers concern science, that they represent original work and real contributions to knowledge, that they had not been published previously, that they be subject to scientific review by peers, that the nominal author be the real author, that they cite previous work, and that they maintain a certain level of politeness toward other researchers. Doubtless there were other such norms, more tacit and less evident to the historian. The point is that for the first time we observe the formal imposition of norms by a scientific institution and by the community of practitioners. The Catholic Church insisted on certain norms in censoring works, too, but the difference is telling. With the Comité de Librairie the community of scientists itself became the agency to articulate and insist that scientific productions meet certain minimum normative standards of the times.

Related to the above, this study has also demonstrated the role of the Academy and its Comité de Librairie as filters through which knowledge had to pass in the process of moving from the private discoveries of individuals to the public sphere where knowledge became available for public consumption.[5] The fact that this was a new filter applied by the established scientific community is the historical novelty to be recognized. With the advent of the Comité de Librairie one does not know whether it became harder to move scientific claims to knowledge into the public realm, but the process certainly became more complicated.

Finally, this study took off from the general understanding that the advent of scientific journals in the 1660s represented a landmark in the history of the scientific press. The *Histoire et Mémoires de l'Académie Royale des Sciences* was indisputably the most important scientific series of the eighteenth century, and in general the Academy and the Comité de Librairie saw to the publication of the largest and most significant body of science and technology appearing anywhere in the world in the eighteenth century. The author's hope is that, having traced in detail the workings of the Comité de Librairie, our awareness of the Academy's *Histoire et Mémoires* and the overall history of the scientific press in the Old Regime is commensurately enhanced. That is, a complicated editorial process was involved, much time and energy were spent, and a lot was at stake, given that the publications of the Academy represented knowledge certified as true and correct by the Academy and the Comité de Librairie. Whatever else may be said in these regards, the author hopes at last to have answered Daniel Roche's question about how papers were chosen for publication, the question that initiated this inquiry.

This account reaches a chronological conclusion with the closing of the Académie Royale des Sciences in 1793. On August 8th of that year the revolutionary Convention voted to eliminate "all academies and literary societies established or endowed by the nation," and so ends not only the history of the Old-Regime Academy of Sciences but also the story of its Comité de Librairie.[6] As much as the closing of the Academy represents the natural terminus for this study of the Comité de Librairie, the movement toward specialist control in the world of science and scientific publishing continued and expanded in the succeeding Académie des Sciences of the Institut de France (1795). That further expansion of control by scientific specialists is the central theme of Maurice Crosland's authoritative 1992 history of the Academy in the succeeding period of its history, *Science Under Control: The French Academy of Sciences, 1795–1914.*[7] The nineteenth-century French Academy of Sciences remained Europe's most luminous scientific institution at least through the first half of the nineteenth century, and the institution stayed at the cutting edge in scientific publishing (and improved upon its Old-Regime predecessor) by initiating its weekly scientific series, the famous *Comptes rendus hebdomadaires des séances* in 1835.

Several other factors altered the shape of science and scientific publishing in the decades that followed. The appearance of disciplinary journals, standardized norms for citation, increased speed of publication, improved mechanisms for distributing journals, and modernized technologies for printing and reproducing visual and graphical materials— these and related novelties resulted in the truly modern scientific press.[8]

Figure 8: An unsigned silver *jeton* of the Académie Royale des Sciences. Obverse: Louis XVI facing right (Feuardent bust #387) with the legend, "LUD.XVI.REX CHRISTIANISS"—"The Most Christian King Louis XVI." Reverse: Minerva seated and facing right surrounded by scientific instruments. Legend is the motto of the Academy, "INVENIT ET PERFICIT"—"Invent and Perfect"; exergue: REGIA SCIENTIARUM ACADEMIA—"The Royal Academy of Sciences." Feuardent #4409var.

Today, in the weekly publication of *Science* or *Nature*, for example, scientific specialists exercise the strictest vigilance over the research that finds its way into print, and they form part of the elaborate social, institutional, and professional enterprise that constitutes science today. The present study has shown that the historical antecedents of this aspect of modern science are to be found in Old-Regime France, in the Académie Royale des Sciences (Fig. 8), and in its now long-defunct Comité de Librairie.

Notes

1. On these points, see McClellan, *Science Reorganized*, Epilogue. Fox and Weisz, *The Organization of Science*, is to be cited in connection with French science in the nineteenth century, and still valuable in these regards is Sharlin, *The Convergent Century*.

2. See Kuhn, "Mathematical versus Experimental Traditions," on this point.

3. Historians of the Paris Academy and contemporary French science have long recognized this quid pro quo relation between state and institution; this is a central theme in Gillispie, *Science and Polity*; see esp. 549–52; see also Hahn, *Anatomy*, 6–10; McClellan, *Science Reorganized*, 25–26.

4. This theme informs Hahn's *Anatomy* throughout, but see especially chapts. 5 and 6.

5. See above, Chapter 1 at notes 8 and 16. Rudwick, "Charles Darwin in London," provides a nuanced perspective on making knowledge public from the point of view of the individual researcher.

6. Hahn, *Anatomy*, chapt. 8 tells this standard story.

7. See especially chapter 12. See also Crosland's "Scientific Credentials" and his "Assessment by Peers."

8. On these points, see the horribly mistitled Hunter, ed., *Thornton and Tully*, 4th ed., and the still valuable Thornton and Tully, *Scientific Books, Libraries and Collectors*, 3rd ed. See also Kronick, *A History of Scientific and Technical Periodicals* and his *Scientific and Technical Periodicals*; Meadows, *Communication in Science*, pp. 72–79.

Appendices

The following statistics (Apppendix 1) and list of members (Appendix 2) of the Comité de Librairie are based on the data compiled by Mme Pierre Gauja and Mme Christiane Demeulenaere-Douyère, past archivists of the Académie des Sciences in Paris. (See their "Comité de Librairie d'après le Règlement de 1731," AdS, DG 31.) The Gauja-Demeulenaere template displays the membership on the Comité by year and by seat or category of membership, that is, ex officio (as president, vice-president, director, sub-director, secretary, or treasurer); permanent member in the classes of the mathematical and physical sciences; annual member; or substitute for an annual or permanent member serving for a year as an officer of the Academy.

The Gauja-Demeulenaere template provides reliable and comprehensive information about the membership of the Comité from 1731, the year that formal statutes regularized the membership and structure of the Comité, through 1793, the year the Convention closed the Academy. (On the regulations of 1731, see above, Chapter 3.) A few lacunae and uncertainties should be noted, however. Annual members for 1775 and 1776 remain unknown. Similarly, we do not know who, if anyone at all, served as permanent or annual members in the tumultuous years, 1791, 1792, or 1793. Not all substitutes are accounted for: for example, Trudaine de Montigny is listed as an annual member and as vice-president for 1772, with no replacement indicated for him as annual member. It is not clear whether substitutions were made on the two occasions when the secretary or treasurer also served as director or sub-director, as secretary Grandjean de Fouchy did in turn in 1769 and 1770 and as Tillet, the assistant treasurer, did in 1778 and 1779. Finally in this connection, when Tillet was assistant treasurer to Buffon (1773–1791) or Condorcet assistant secretary to Grandjean de Fouchy

(1772–1776), the record does not indicate clearly whether the primary officer or the assistant or both sat on the Comité. These and other minor uncertainties are indicated in the notes to these appendices. Additional information will modify somewhat the statistics given for the Comité de Librairie in Appendix 1 or the list of members in Appendix 2. Still, the data compiled by Mmes Gauja and Demeulenaere more than suffice to gain an accurate general overview of the membership of the Comité de Librairie.

Information concerning the rank of Comité members in the Academy (*honoraire, pensionnaire*, etc.) at the time of service on the Comité is drawn from the Academy's *Index biographique*. The same source was used to identify the scientific sections and disciplinary specialities from which *pensionnaires* and *associés* were drawn.

Appendix 1.
Membership Statistics for the Comité de Librairie: 1731–1793

a. Total number of members, 1731–1793:   90

b. Average number of years of service: 6.6 (7.0)[1]

c. Seat on Comité:

	Number of individuals	Average Years on Comité
President/Vice-Presidents	34	3.7
Director/Sub-Directors	40	3.0
Secretary/Treasurers	8	18 (23.7)[2]
Permanent members	8	15.0
Annual members	34	3.3
Substitutes	9	1.4

d. Rank in Academy:

	Number	Percentage	Average Years on Comité
Honoraire	32	35.6	3.9
Secretary/Treasurer	6[3]	6.7 (8.9)[4]	18 (23.7)[5]
Pensionnaire	47	52.2	7.5
Associé	5	5.6	2.2

e. Scientific Sections represented by *pensionnaires* and *associés* on the
Comité de Librairie:[6]

Number of Members from:

Mathematical Sciences (26):
Geometry	9
Mechanics	11
Astronomy	6

Physical Sciences (26):
Anatomy	5
Chemistry	11
Botany/Natural History	10

Appendix 2.
Members of the Comité de Librairie: 1731–1793

d'Aguesseau, Henri-François (1668–1751)
>*Honoraire* from 1728.
>Service on Comité: Ex officio as president, 1739
>Years on Comité: 1

d'Aiguillon, Armand-Louis Duplessis de Richelieu, duke
(1683–1750)
>*Honoraire* from 1744.
>Service on Comité:
>>Ex officio as vice-president, 1746
>>Ex officio as president, 1747
>Years on Comité: 2

d'Alembert, Jean le Rond (1717–1783)
>*Adjoint* in 1741; *pensionnaire* in mechanics from 1765.
>Service on Comité:
>>Annual member, 1767
>>Ex officio as sub-director, 1768
>>Ex officio as director, 1769
>>Permanent member, mathematical sciences, 1772–1783
>Years on Comité: 15

Amelot, Antoine-Jean (?–1794)
  Honoraire from 1777.
  Service on Comité:
      Ex officio as vice-president, 1778
      Ex officio as president, 1779
  Years on Comité: 2

Amelot, Jean-Jacques (1689–1749)
  Honoraire from 1741.
  Service on Comité:
      Ex officio as vice-president, 1743, 1748
      Ex officio as president, 1744, 1749
  Years on Comité: 4

Arcy, Patrick, count d' (1725–1779)
  Adjoint in 1749; *pensionnaire* in geometry from 1771.
  Service on Comité:
      Ex officio as sub-director, 1774
      Ex officio as director, 1775
  Years on Comité: 2

Argenson, Marc-Pierre de Voyer de Paulmy, count d' (1696–1764)
  Honoraire from 1726.
  Service on Comité:
      Ex officio as vice-president, 1737, 1738, 1740, 1742,
          1753
      Ex officio as president, 1731, 1741, 1754
  Years on Comité: 8

Berthollet, Claude-Louis (1748–1822)
  Adjoint in 1780; *associé* in chemistry and metallurgy from
  1785; *pensionnaire* in chemistry in 1792.
  Service on Comité:
      Annual member, 1786, 1787.
  Years on Comité: 2

Bertin, Henri-Léonard-Jean-Baptiste (1720–1792)
  Honoraire from 1761.
  Service on Comité:
      Ex officio as vice-president, 1756, 1763, 1769
      Ex officio as president, 1764, 1770
  Years on Comité: 5

Bézout, Étienne (1730–1783)
>   *Adjoint* in 1758; supernumerary *pensionnaire* in mechanics
>   from 1779
>   Service on Comité:
>>   Annual member, 1779, 1780
>   Years on Comité: 2

Bignon, Abbé Jean-Paul (1662–1743)
>   *Honoraire* from 1691.
>   Service on Comité:
>>   Ex officio as vice-president, 1731, 1733
>>   Ex officio as president, 1732, 1734
>   Years on Comité: 4

Bochart de Saron, Jean-Baptiste-Gaspard (1730–1794)
>   *Honoraire* from 1779.
>   Service on Comité:
>>   Ex officio as vice-president, 1782, 1787
>>   Ex officio as president, 1783, 1788
>   Years on Comité: 4

Borda, Jean-Charles de (1733–1799)
>   *Adjoint* in 1756; *pensionnaire* in geometry from 1772.
>   Service on Comité:
>>   Annual Member, 1773, 1774
>>   Ex officio as sub-director, 1776
>>   Ex officio as director, 1777
>>   Permanent member, mathematical sciences, 1784–1790
>   Years on Comité: 11

Bossut, Charles (1730–1814)
>   *Correspondant* in 1753; *adjoint* in 1768; *pensionnaire* in geometry
>   from 1779.
>   Service on Comité:
>>   Ex officio as sub-director, 1783
>>   Ex officio as director, 1784
>>   Annual member, 1785, 1786, 1789, 1790
>   Years on Comité: 6

Bouguer, Pierre (1698–1758)
>   *Associé* in 1731; *pensionnaire* in astronomy from 1735.
>   Service on Comité:
>>   Substitute for Réaumur, 1746
>>   Ex officio as sub-director, 1747, 1754

Ex officio as director, 1748, 1755
Annual member, 1750, 1751, 1756, 1757
Permanent member, mathematical sciences, 1758
Years on Comité: 10

Bourdelin, Louis-Claude (1696–1777)
Adjoint in 1727; *pensionnaire* in chemistry from 1752.
Service on Comité:
Annual member, 1753, 1754, 1757, 1758, 1761, 1762, 1766, 1767
Years on Comité: 8

Breteuil, Louis-August Le Tonnelier, baron de (1730–1807)
Honoraire from 1785.
Service on Comité:
Ex officio as vice-president, 1786
Ex officio as president, 1787
Years on Comité: 2

Brisson, Mathurin-Jacques (1723–1806)
Adjoint in 1759; *associé* in botany from 1779 to 1782; *pensionnaire* in general physics from 1785.
Service on Comité:
Annual member 1780
Years on Comité: 1

Buffon, Georges-Louis Leclerc, count de (1707–1788)
Adjoint in 1734; *associé* in botany in 1739; treasurer from 1774.
Service on Comité:
Ex officio as treasurer, 1744–1773 (possibly 1744–1788)[7]
Years on Comité: 30/45

Cadet, Louis-Claude (1731–1799)
Adjoint in 1766; *pensionnaire* in chemistry from 1777.
Service on Comité:
Ex officio as sub-director, 1782
Ex officio as director, 1783
Years on Comité: 2

Camus, Abbé Charles-Étienne-Louis (1699–1768)
Adjoint in 1727; *pensionnaire* in geometry from 1741.
Service on Comité:
Annual member, 1742, 1743, 1746, 1747, 1752, 1753, 1763, 1764

Ex officio as sub-director, 1749, 1760
Ex officio as director, 1750, 1761
Replacement for de Mairan, 1759
Years on Comité: 14

Cassini, Jacques [Cassini II] (1677–1756)
Associé in 1699; *pensionnaire* in astronomy from 1712.
Service on Comité:
Ex officio as sub-director, 1731, 1738
Ex officio as director, 1732, 1739
Annual member, 1733, 1734, 1737, 1740
Years on Comité: 8

Cassini de Thury, César-François [Cassini III] (1714–1784)
Adjoint in 1735; *pensionnaire* in astronomy from 1745.
Service on Comité:
Annual member, 1754, 1755
Ex officio as sub-director, 1757, 1766, 1770
Ex officio as director, 1758, 1767, 1771
Years on Comité: 8

de Castries, Charles-Eugène-Gabriel de La Croix, marquis (1727–1801)
Honoraire from 1788.
Service on Comité:
Ex officio as vice-president, 1790
Ex officio as president, 1791
Years on Comité: 2

Chaulnes, Michel-Ferinand d'Albert d'Ailly, duke de (1714–1769)
Honoraire from 1743.
Service on Comité:
Ex officio as vice-president, 1745, 1749, 1758
Ex officio as president, 1746, 1750, 1759
Years on Comité: 6

Clairaut, Alexis-Claude (1713–1765)
Adjoint in 1731; *pensionnaire* in mechanics from 1738.
Service on Comité:
Substitute for Réaumur, 1739
Substitute for de Mairan, 1760
Annual member, 1741, 1761, 1762
Ex officio as sub-director, 1743, 1756, 1765
Ex officio as director, 1744, 1757
Years on Comité: 10

Condorcet, Marie-Jean-Antoine-Nicolas Caritat, marquis de
(1743–1794)
> *Adjoint* in 1769; supernumerary *pensionnaire* in mechanics and
> assistant secretary from 1773; permanent secretary from 1776.
> Service on Comité:
>> Ex officio: assistant secretary, 1773–1776
>> Ex officio: permanent secretary, 1776–1793
> Years on Comité: 18/21[8]

Coulomb, Charles-Augustin (1736–1806)
> *Correspondant* in 1774; *adjoint* in 1781; *associé* in mechanics from
> 1784.
> Service on Comité
>> Annual member, 1787, 1788
> Years on Comité: 2

Couplet, Pierre (?–1743)
> *Élève* in mechanics in 1696; *adjoint* in mechanics in 1716; treasurer
> from 1717.
> Service on Comité:
>> Ex officio as treasurer, 1731–1743
> Years on Comité: 13

Courtanvaux, François-César Le Tellier, marquis de (1718–1781)
> *Honoraire* from 1765.
> Service on Comité:
>> Ex officio as vice-president, 1768, 1774
>> Ex officio as president, 1769, 1775
> Years on Comité: 4

Darcet, Jean (1725–1804)
> *Associé* in 1784; *pensionnaire* in natural history and mineralogy
> from 1786.
> Service on Comité:
>> Ex officio as sub-director, 1792
>> Ex officio as president, 1793
> Years on Comité: 2

Daubenton, Louis-Jean-Marie (1716–1800)
> *Adjoint* in 1744; *pensionnaire* in anatomy from 1760.
> Service on Comité:
>> Annual member, 1773, 1774, 1777
> Years on Comité: 3

Desmarest, Nicolas (1725–1815)
> *Adjoint* in 1771; *pensionnaire* in mechanics from 1782; *pensionnaire* in natural history and mineralogy from 1785.
> Service on Comité:
>> Ex officio as sub-director, 1785
>> Ex officio as director, 1786
> Years on Comité: 2

Dufay, Charles-François de Cisternay (1698–1739)
> *Adjoint* in 1723; *pensionnaire* in chemistry from 1731.
> Service on Comité:
>> Ex officio as sub-director, 1732, 1737
>> Ex officio as director, 1733, 1738
>> Annual member, 1735, 1736
> Years on Comité: 6

Duhamel du Monceau, Henri-Louis (1700–1782)
> *Adjoint* in 1728; *pensionnaire* in botany from 1738.
> Service on Comité:
>> Ex officio as sub-director, 1742, 1755, 1767
>> Ex officio as director, 1743, 1756, 1768
>> Permanent member, physical sciences, 1744–1782
> Years on Comité: 41

Fleury, André-Hercule, cardinal de (1653–1743)
> *Honoraire* from 1721.
> Service on Comité:
>> Ex officio as president, 1738.
> Years on Comité: 1

Fontenelle, Bernard Le Bovier de (1657–1757)
> Academician in 1697 in geometry; secretary from 1697.
> Service on Comité:
>> Ex officio as permanent secretary, 1731–1740.
> Years on Comité: 10.

Fouchy, Jean-Paul Grandjean de (1707–1788)
> *Adjoint* in 1731; *associé* in astronomy in 1741; secretary from 1743.
> Service on Comité:
>> Ex officio as permanent secretary, 1743–1776
>> Ex officio as sub-director, 1769
>> Ex officio as director, 1770
> Years on Comité: 34

Fougeroux de Bondaroy, August-Denis (1732–1789)
    Adjoint in 1758; *pensionnaire* in botany from 1779.
    Service on Comité:
        Ex officio as sub-director, 1786
        Ex officio as director, 1787
    Years on Comité: 2

Geoffroy, Claude-Joseph [Geoffroy *cadet*] (1685–1752)
    Élève in 1707; *pensionnaire* in chemistry from 1723.
    Service on Comité:
        Annual member, 1731, 1741
        Substitute for Réaumur, 1740
    Years on Comité: 3

Hellot, Jean (1685–1766)
    Adjoint in 1735; *pensionnaire* in chemistry from 1739.
    Service on Comité:
        Annual member, 1748, 1749, 1752, 1765
        Ex officio as sub-director, 1750, 1763
        Ex officio as director, 1751, 1764
        Replacement for Duhamel du Monceau, 1755, 1756
    Years on Comité: 10

Jeaurat, Edme-Sébastien (1725–1803)
    Adjoint in 1763; *pensionnaire* in geometry from 1783.
    Service on Comité:
        Ex officio as sub-director, 1791
        Ex officio as director, 1792
    Years on Comité: 2

Jussieu, Antoine de (1686–1758)
    Élève in 1712; *pensionnaire* in botany from 1715.
    Service on Comité:
        Annual member, 1733, 1734, 1737, 1738
        Years on Comité: 4

Jussieu, Antoine Laurent de (1748–1836)
    Adjoint in 1773; *associé* in botany in 1782; *pensionnaire* in botany
        and agriculture from 1785.
    Service on Comité:
        Ex officio as sub-director, 1790
        Ex officio as director, 1791
    Years on Comité: 2

Jussieu, Bernard de (1699–1777)
>   *Adjoint* in 1725; *pensionnaire* in botany from 1739.
>   Service on Comité[9]:
>   >   Annual member, 1739[10], 1744, 1745, 1759, 1760, 1763,
>   >   1764, 1770, 1771, 1772
>   >   Substitute for Duhamel du Monceau, 1767, 1768
>   >   Ex officio as sub-director, 1753
>   >   Ex officio as director, 1754
>   Years on Comité: 14

La Condamine, Charles-Marie de (1701–1774)
>   *Adjoint* in 1730; *pensionnaire* in chemistry from 1739.
>   Service on Comité:
>   >   Annual member, 1746, 1747
>   >   Ex officio as sub-director, 1748
>   >   Ex officio as director, 1749
>   Years on Comité: 4

Lagrange, Joseph-Louis de (1736–1813)
>   *Associé étranger* in 1772; *pensionnaire* (veteran) from 1787.[11]
>   Service on Comité:
>   >   Ex officio as director, 1788
>   Years on Comité: 1

La Lande, Joseph-Jérome Le François (1732–1807)
>   *Adjoint* in 1753; *pensionnaire* in astronomy from 1772.
>   Service on Comité:
>   >   Ex officio as sub-director, 1781
>   >   Ex officio as director, 1782
>   Years on Comité: 2

La Luzerne, César-Henri, count de (1737–1799)
>   *Honoraire* from 1788.
>   Service on Comité:
>   >   Ex officio as vice-president, 1789
>   >   Ex officio as president, 1790
>   Years on Comité: 2

Laplace, Pierre-Simon de (1749–1812)
>   *Adjoint* in 1773; *pensionnaire* in mechanics from 1785.
>   Service on Comité:
>   >   Ex officio as vice-president, 1793
>   Years on Comité: 1

La Rochefoucault d'Anville, Louis-Alexandre, duke de (1743–1792)
:   *Honoraire* from 1781.
    Service on Comité:
        Ex officio as vice-president, 1783, 1792
        Ex officio as president, 1784
    Years on Comité: 3

Lavoisier, Antoine-Laurent de (1743–1794)
:   *Adjoint* in 1768; *pensionnaire* in chemistry from 1778; treasurer from 1791.
    Service on Comité:
        Annual member, 1778, 1779
        Permanent member, physical sciences, 1783–1790
        Ex officio as sub-director, 1784
        Ex officio as director, 1785
        Treasurer, 1791–1793
    Years on Comité: 13

La Vrillière, Louis Phélypeaux, count de Saint-Florentin, duke de (1705–1777)
:   *Honoraire* from 1740.
    Service on Comité:
        Ex officio as vice-president, 1741, 1747, 1754, 1761, 1767, 1773
        Ex officio as president, 1742, 1748, 1755, 1756, 1762, 1768, 1774
    Years on Comité: 13

Le Gentil, Guillaume-Joseph-Hyacinthe-Jean-Baptiste (1725–1792)
:   *Adjoint* in 1753; supernumerary *pensionnaire* in astronomy from 1782.
    Service on Comité:
        Ex officio as sub-director, 1789
        Ex officio as director, 1790
    Years on Comité: 2

Lémery, Louis (1677–1743)
:   *Élève* in 1700; *pensionnaire* in chemistry from 1715.
    Service on Comité:
        Permanent member, physical sciences, 1731–1743
    Years on Comité: 13

Le Monnier, Pierre-Charles (1715–1799)
>   *Adjoint* in 1736; *pensionnaire* in astronomy from 1746.
>   Service on Comité:
>>   Ex officio as sub-director, 1751, 1764
>>   Ex officio as director, 1752, 1765
>>   Annual member, 1790[12]
>   Years on Comité: 5

Le Roy, Jean-Baptiste (1719–1800)
>   *Adjoint* in 1751; *pensionnaire* in mechanics from 1770; *pensionnaire* in general physics from 1785.
>   Service on Comité:
>>   Ex officio as sub-director, 1772, 1777
>>   Ex officio as director, 1773, 1778
>   Years on Comité: 4

Loménie de Brienne, Étienne-Charles de (1727–1794)
>   *Honoraire* from 1787.
>   Service on Comité:
>>   Ex officio as vice-president, 1788
>>   Ex officio as president, 1789
>   Years on Comité: 2

Lowendal, Ulric-Frédéric-Woldemar, count de (1700–1755)
>   *Honoraire* from 1754.
>   Service on Comité:
>>   Ex officio as vice-president, 1755
>   Years on Comité: 1

Luynes, Paul d'Albert, cardinal de (1703–1788)
>   *Honoraire* from 1755.
>   Service on Comité:
>>   Ex officio as vice-president, 1757
>>   Ex officio as president, 1758
>   Years on Comité: 2

Macquer, Pierre-Joseph (1718-1784)
>   *Adjoint* in 1745; *pensionnaire* in chemistry from 1772.
>   Service on Comité:
>>   Ex officio as sub-director, 1773
>>   Ex officio as director, 1774
>>   Annual member, 1784
>   Years on Comité: 3

Maillebois, Yves-Marie Desmarets, count, then marquis de (1715–1791)
>   *Honoraire* from 1749.
>   Service on Comité:
>>   Ex officio as vice-president, 1750, 1770, 1775, 1781, 1785
>>   Ex officio as president, 1751, 1771, 1776, 1782, 1786
>   Years on Comité: 10

Mairan, Jean-Baptiste Dortous de (1678–1771)
>   *Associé* in 1718; *pensionnaire* in geometry from 1719; secretary, 1741–1743.
>   Service on Comité:
>>   Substitute for Réaumur 1731, 1735, 1747
>>   Ex officio as sub-director, 1736, 1744, 1759
>>   Ex officio as director, 1737, 1745, 1760
>>   Annual member, 1739, 1740, 1748, 1749, 1758
>>   Ex officio as permanent secretary, 1741–1743
>>   Permanent member, mathematical sciences, 1759-1771
>   Years on Comité: 28

Malesherbes, Chrétien-Guillaume de Lamoignon de (1721–1794)
>   *Honoraire* from 1750.
>   Service on Comité:
>>   Ex officio as vice-president, 1751, 1759, 1764, 1780
>>   Ex officio as president, 1752, 1760, 1765, 1781
>   Years on Comité: 8

Malouin, Paul-Jacques (1701–1777)
>   *Adjoint* in 1742; *pensionnaire* in chemistry from 1766.
>   Service on Comité:
>>   Ex officio as sub-director, 1771
>>   Ex officio as director, 1772
>   Years on Comité: 2

Marchant, Jean (?–1738)
>   Academician in 1678; *pensionnaire* in botany from 1699.
>   Service on Comité:
>>   Annual member, 1732
>>   Years on Comité: 1

Maupertuis, Pierre-Louis Moreau de (1698–1759)
> *Adjoint* in 1723; *pensionnaire* in geometry from 1731.
> Service on Comité:
>> Substitute for Réaumur (sub-director), 1734
>> Ex officio as sub-director, 1735, 1741
>> Ex officio as director, 1736, 1742
> Years on Comité: 5

Maurepas, Jean-Frédéric Phélypeaux de Pontchartrain, count de (1701–1781)
> *Honoraire* from 1725.
> Service on Comité:
>> Ex officio as vice-president, 1735, 1736, 1739
>> Ex officio as president, 1737, 1740
> Years on Comité: 5

Montigny, Étienne Mignot de (1714–1782)
> *Adjoint* in 1740; *pensionnaire* in mechanics from 1758.
> Service on Comité[13]:
>> Annual member, 1759, 1760, 1765, 1766, 1777, 1778, 1781
>> Ex officio as sub-director, 1762, 1779
>> Ex officio as director, 1763, 1780
> Years on Comité: 11

Montmirail, Charles-François Le Tellier, marquis de (1734–1764)
> *Honoraire* from 1761.
> Service on Comité:
>> Ex officio as vice-president, 1762
>> Ex officio as president, 1763
> Years on Comité: 2

Morand, Saveur-François (1697–1773)
> *Adjoint* in 1722; *pensionnaire* in anatomy from 1741.
> Service on Comité:
>> Annual member, 1742, 1743, 1750, 1751, 1755, 1756, 1768, 1769
>> Ex officio as director, 1745, 1746, 1759, 1781
>> Ex officio as sub-director, 1758, 1766, 1780
> Years on Comité: 15

Nicole, François (1683–1758)[14]
>*Élève* in 1707; *pensionnaire* in mechanics from 1724.
>Service on Comité:
>>Ex officio as sub-director, 1733, 1740
>>Ex officio as director, 1734, 1741
>>Annual member, 1731, 1732, 1735, 1736, 1738, 1744, 1745
>>Replacement for Réaumur, 1752, 1753
>Years on Comité: 13

Noailles, Jean-Paul-François, duke de (1739–1824)
>*Honoraire* from 1777.
>Service on Comité:
>>Ex officio as vice-president, 1779, 1784, 1791
>>Ex officio as president, 1780, 1785, 1792
>Years on Comité: 6

Nollet, Abbé Jean-Antoine (1700–1770)
>*Adjoint* in 1739; *pensionnaire* in mechanics from 1757.
>Service on Comité:
>>Ex officio as sub-director, 1761, 1770
>>Ex officio as director, 1762
>>Annual member, 1768, 1769
>Years on Comité: 5

Paulmy d'Argenson, Marc-Antoine-René de Voyer, marquis de (1722–1787)
>*Honoraire* from 1764.
>Service on Comité:
>>Ex officio as vice president, 1766, 1771, 1777
>>Ex officio as president, 1767, 1772, 1778
>Years on Comité: 6

Polignac, Melchior, cardinal de (1661–1741)
>*Honoraire* from 1715.
>Service on Comité:
>>Ex officio as president, 1733
>Years on Comité: 1

Portal, Antoine (1742–1832)
>*Adjoint* in 1769; *associé* in anatomy from 1774; *pensionnaire* in anatomy from 1784.

Service on Comité:
>   Annual member, 1782, 1783
> Years on Comité: 2

Réaumur, René-Antoine Ferchault de (1683–1757)
> *Élève* in 1708; *pensionnaire* in mechanics from 1711.
> Service on Comité:
>>   Permanent member, mathematical sciences, 1731–1757
>>   Ex officio as sub-director, 1734, 1739, 1746, 1752,
>>   Ex officio as director, 1735, 1740, 1747, 1753
> Years on Comité: 27

Richelieu, Louis-François-Armand Duplessis, duke de (1696–1788)
> *Honoraire* from 1731.
> Service on Comité:
>>   Ex officio as vice-president, 1734
>>   Ex officio as president, 1735
> Years on Comité: 2

Rochon, Abbé Alexis-Marie de (1741–1817)
> *Correspondant* in 1767; *adjoint* in 1771; *pensionnaire* in mechanics
>   from 1783.
> Service on Comité:
>>   Annual member, 1783, 1784
>>   Ex officio as sub-director, 1787
> Years on Comité: 3

Rouillé, Antoine-Louis, count de Jouy (1689–1761)
> *Honoraire* from 1751.
> Service on Comité:
>>   Ex officio as vice-president, 1752
>>   Ex officio as president, 1753
> Years on Comité: 2

Sabatier, Raphaël-Bienvenu (1732–1811)
> *Adjoint* in 1773; *associé* in anatomy from 1784.
> Service on Comité:
>>   Annual member, 1785, 1788, 1789
> Years on Comité: 3

Sage, Balthazar-Georges (1740–1824)
> *Adjoint* in 1770; *pensionnaire* in chemistry in 1784; *pensionnaire* in
>   natural history and mineralogy from 1785.

Service on Comité:
>   Ex officio as sub-director, 1788
>   Ex officio as director, 1789

Years on Comité: 2

Séchelles, Jean Moreau de (1690–1760)
>   *Honoraire* from 1755.

Service on Comité:
>   Ex officio as president, 1757

Years on Comité: 1

Tenon, Jacques-René (1724–1816)
>   *Adjoint* in 1759; *pensionnaire* in anatomy from 1773.

Service on Comité:
>   Ex officio as sub-director, 1775
>   Ex officio as director, 1776

Years on Comité: 2

Tillet, Mathieu (1714–1791)
>   *Adjoint* in botany from 1758; supernumerary *pensionnaire* from 1772; assistant treasurer, 1772–1787; treasurer, 1788–1791.

Service on Comité:
>   Ex officio as assistant treasurer, 1772–1787
>   Ex officio as treasurer, 1788–1790
>   Ex officio as sub-director, 1778
>   Ex officio as director, 1779

Years on Comité: 3/19[15]

Torcy, Jean-Baptiste Colbert, marquis de (1665–1746)
>   *Honoraire* from 1718.

Service on Comité:
>   Ex officio as vice-president, 1732
>   Ex officio as president, 1736, 1743

Years on Comité: 3

Trudaine, Daniel-Charles (1703–1769)
>   *Honoraire* from 1743.

Service on Comité:
>   Ex officio as vice-president, 1744, 1760
>   Ex officio as president, 1745, 1761

Years on Comité: 4

Trudaine de Montigny, Jean-Charles-Philibert (1733–1777)
> *Honoraire* from 1764.
> Service on Comité[16]:
>> Ex officio as vice-president, 1765, 1772, 1776
>> Ex officio as president, 1766, 1773, 1777
>> Annual member, 1770, 1771, 1772
> Years on Comité: 8

Vandermonde, Alexandre (1735–1796)
> *Adjoint* in 1771; *associé* in geometry from 1779; *pensionnaire* in geometry from 1785.
> Service on Comité:
>> Annual member, 1781, 1782
> Years on Comité: 2

Notes

1. Two sets of figures are given, depending on whether one counts Buffon as serving on the Comité de Librairie as treasurer all the way to 1788 and whether Tillet (as assistant treasurer) and Condorcet (assistant secretary) are counted as overlapping with their full-fledged counterparts; see below at notes 7, 8, and 15.

2. The parenthetical number counts Buffon as serving on the Comité de Librairie as treasurer through 1788 and Tillet and Condorcet (as assistant treasurer and assistant secretary, respectively) serving concurrently with the treasurer and permanent secretary.

3. This number excludes two individuals: (1) Mairan who served as secretary from 1741 to 1743, but otherwise was a *pensionnaire* in the Academy for over fifty years and here is counted as such; (2) Lavoisier, who was treasurer from 1791 to 1793, but who was a *pensionnaire* from 1778 and for present purposes is likewise included in that group.

4. The percentage in parenthesis includes the years served by Mairan as permanent secretary and by Lavoisier as treasurer; see previous note.

5. See note 2 above.

6. See Hahn, *Anatomy*, 98–101; Gillispie, *Science and Polity*, 91, and above, Chapt. 3. In this tabulation the three people from natural history who served on the Comité de Librairie in the period 1785 to 1793 (Darcet, Desmarest, and Sage) have been lumped with botany. Technically, in the reform of 1785, chemistry became "chemistry and metallurgy," botany became "botany and agriculture," and natural history was "natural history and mineralogy."

7. Buffon was treasurer of the Academy from 1744 until his death in 1788. Mathieu Tillet was named assistant treasurer at the end of 1772. As noted above, it is not clear whether one or both individuals sat on the Comité in the period from 1773 through 1788. Therefore, two ranges of dates are given.

8. Condorcet was named assistant secretary in 1773 to relieve Grandjean de Fouchy, the permanent secretary. Condorcet assumed the responsibilities of permanent secretary in 1776. This instance is thus similar to the Buffon-Tillet case mentioned in the previous note. Condorcet seems to have taken over the minutes of the meetings of the Comité de Librairie after August 1776, so it may be that he joined the Comité at that point; see RCL II/63ff (August 1776). It is still unclear, however, whether Condorcet sat on the Comité de Librairie as assistant secretary in the period 1773–1776, and hence, again, two spans of years are given.

9. Regarding the de Jussieu annual memberships from 1759 and the substitute positions in 1767 and 1768, the Gauja/Demeulenaere template does not differentiate Bernard de Jussieu from his brother, Joseph de Jussieu (1704–1779). But Joseph de Jussieu became a veteran *associé* of the Academy in 1758, so it seems reasonable to list Bernard de Jussieu as filling these slots in these years.

10. Possibly Antoine de Jussieu.

11. For the purposes of Appendix 1 above, Lagrange is counted as a *pensionnaire* geometer.

12. Conceivably his brother, Louis-Guillaume Le Monnier (1717–1799).

13. There may be possible confusions in dates between this Montigny and Trudaine de Montigny; see below at note 16.

14. On the Gauja/Demeulenaere template a certain Nisle is listed as an annual member for 1731 and 1732. This is a typographical error for Nicole.

15. Similar to the cases of Buffon and Condorcet (above at notes 7 and 8), it is not known whether Tillet served on the Comité de Librairie as assistant treasurer in the years 1773–1788. Hence, again, two figures are given.

16. The dates Trudaine de Montigny served on the Comité de Librairie may possibly be confused with those of Mignot de Montigny; see above at note 13.

Bibliography

Académie des Sciences. *Index Biographique des Membres et Correspondants de l'Académie des Sciences*. Paris: Gauthier-Villars, 1979.

Aucoc, Léon. *L'Institut de France: Lois, Statuts et Règlements concernant les anciennes académies et l'Institut de 1635 à 1889*. Paris: Imprimerie Nationale, 1889.

Baker, John R. "Trembley, Abraham," in Gillispie, ed. *Dictionary of Scientific Biography*, vol. 13, 457–458. New York: Charles Scribner's Sons, 1976.

Baker, Keith M. *Condorcet: From Natural Philosophy to Social Mathematics*. Chicago and London: University of Chicago Press, 1975.

———. "Les Débuts de Condorcet au Secrétariat de l'Académie Royale des Sciences (1773–1776)." *Revue d'Histoire des Sciences* 20 (1967): 229–280.

Bazerman, Charles. *Shaping Written Knowledge: The Genre and Activity of the Experimental Article in Science*. Madison: University of Wisconsin Press, 1988.

Bénézit, Émmanuel. *Dictionnaire critique et documentaire des peintres sculpteurs, dessinateurs et graveurs*, 10 vols. Paris: Librairie Gründ, 1976.

Blay, Michel and Robert Halleux, eds. *La Science Classique XVIᵉ–XVIIIᵉ siècle: Dictionnaire critique*. Paris: Flammarion, 1998.

Brian, Éric. "L'Académie royale des science de l'absolutisme à la Révolution," 15–32, in Brian and Demeulenaere-Doyère, eds., *Histoire et mémoire*.

Brian, Éric and Christiane Demeulenaere-Doyère, eds. *Actes du Colloque, «Règlement, usages et science dans la France de l'Absolutisme.»* Paris: Tec & Doc Lavoisier, 2002.

Brian, Éric and Christiane Demeulenaere-Douyère, eds. *Histoire et mémoire de l'Académie des sciences: Guide de recherches*. Paris: Tec & Doc Lavoisier, 1996.

Cohen, H. Floris. *The Scientific Revolution: A Historiographical Inquiry*. Chicago: University of Chicago Press, 1994.

Conant, James B. "The Overthrow of Phlogiston Theory: The Chemical Revolution of 1775–1789," 65–116, in Conant, ed., *Harvard Case Studies*, vol. 1.

Conant, James B., ed. *Harvard Case Studies in Experimental Science*, 2 vols. Cambridge, MA: Harvard University Press, 1952.

Crépel, Pierre. "Une curieuse lettre de Borda à Condorcet," 325–337, in Brian and Demeulenaere-Doyère, eds., *Histoire et mémoire*.

Crosland, Maurice. *Science Under Control: The French Academy of Sciences, 1795–1914*. Cambridge and New York: Cambridge University Press, 1992.

———. "Scientific Credentials: Record of Publications in the Assessment of Qualifications for Election to the French Académie des Sciences." *Minerva* 19 (1981): 605–631.

———. "Assessment by Peers in 19th Century France: The Manuscript Report on Candidates for Election to the Académie des sciences." *Minerva* 24 (1986): 413–432.

Daniel, Hans-Dieter. *Guardians of Science: Fairness and Reliability of Peer Review*, William E. Russey, trans. Weinheim, Germany and New York: VCH, 1993.

Darnton, Robert. *The Great Cat Massacre and Other Episodes of French Cultural History*. New York: Basic Books, 1984; Vintage Books, 1985.

———. *The Kiss of Lamourette: Reflections in Cultural History*. New York and London: W. W. Norton, 1990.

———. *Mesmer and the End of the Enlightenment in France*. Cambridge, MA: Harvard University Press, 1970.

Darnton, Robert and Daniel Roche, eds. *Revolution in Print: The Press in France, 1775–1800*. Berkeley: University of California Press, 1989.

Daston, Lorraine. "The Language of Strange Facts in Early Modern Science," 20–38, in Lenoir, ed., *Inscribing Science*.

Dawson, Virginia P. *Nature's Enigma: The Problem of the Polyp in the Letters of Bonnet, Trembley, and Réaumur*. Philadelphia: American Philosophical Society, 1987. [*Memoirs of the American Philosophical Society*, vol. 174.]

Doneaud du Plan, Alfred. *Histoire de l'Académie de Marine*. 6 parts. Paris: Berger-Levrault, 1879–1882.

Duprat, Gabrielle. "Les Dessinateurs d'Histoire Naturelle en France au 18ᵉ siècle," 451–470 in Lawrence, ed., *Adanson*, vol. 2.

Eamon, William. *Science and the Secrets of Nature: Books of Secrets Medieval and Early Modern Culture*. Princeton, NJ: Princeton University Press, 1994.

Feuardent, Félix. *Jetons et méreaux depuis Louis XI jusqu'à la fin du Consulat de Bonaparte*, 3 vols. Paris: Rollin et Feuardent, 1904–1915. Reprint, 4 vols.: Paris: Maison Platt, 1995.

Ford, Brian J. "Eighteenth-Century Scientific Publishing," 216–257, in Hunter, ed., *Thornton and Tully*.

Fox, Robert and George Weisz, eds. *The Organization of Science and Technology in France, 1808–1914*. Cambridge: Cambridge University Press; Paris: Éditions de la Maison des Sciences de l'Homme, 1980.

Gascoigne, Robert Mortimer. *A Historical Catalogue of Scientific Periodicals, 1665–1900: With a Survey of Their Development*. New York and London: Garland, 1985.

Gillispie, Charles C. *Science and Polity in France at the End of the Old Regime*. Princeton: Princeton University Press, 1980.

Golinski, Jan. *Science as Public Culture: Chemistry and Enlightenment in Britain, 1760–1820*. Cambridge: Cambridge University Press, 1992.

Grafton, Antony. *The Footnote: A Curious History*. Cambridge, MA: Harvard University Press, 1997.

Greenberg, John L. *The Problem of the Earth's Shape from Newton to Clairaut: The Rise of Mathematical Science in Eighteenth-Century Paris and the Fall of "Normal" Science*. Cambridge and New York: Cambridge University Press, 1995.

Guénoun, Anne-Sylvie. "Les publications de l'Académie des sciences," 107–140, in Brian and Demeulenaere-Doyère, eds., *Histoire et mémoire*.

Gumbrecht, Hans Ulrich and K. Ludwig Pfeiffer, eds. *Materialities of Communication*, William Whobrey, trans. Stanford, CA: Stanford University Press, 1994.

Hahn, Roger. *The Anatomy of a Scientific Institution: The Paris Academy of Sciences, 1666–1803*. Berkeley: University of California Press, 1971.

Hall, A. R. *Philosophers at War: The Quarrel between Newton and Leibniz*. Cambridge: Cambridge University Press, 1980.

Halleux, Robert, James McClellan, Daniela Berariu, and Geneviève Xhayet. *Les publications de l'Académie royale des sciences de Paris (1666–1793)*. 2 vols. Turnhout, Belgium: Brepols, 2001.

Halliday, M. A. K. and J. R. Martin. *Writing Science: Literacy and Discursive Power*. Pittsburgh and London: University of Pittsburgh Press, 1993.

Heilbron, J[ohn] L. *Electricity in the 17th and 18th Centuries*. Berkeley: University of California Press, 1979.

———. "Nollet, Jean-Antoine," in Charles Gillispie, ed., *Dictionary of Scientific Biography*, vol. X, 145–148. New York: Charles Scribner's Sons, 1974.

Hessenbruch, Arne, ed. *Reader's Guide to the History of Science*. London, Chicago: Fitzroy Dearborn, 2000.

Hunter, Andrew, ed. *Thornton and Tully's Scientific Books, Libraries, and Collectors, Fourth Edition*. Aldershot: Ashgate, 2000.

Jacobi, Daniel. *La communication scientifique: Discours, figures, modèles*. Grenoble: Presses Universitaires de Grenoble. 1999.

———. *Diffusion et vulgarisation: Itinéraires du texte scientifique*. [Annales littéraires de l'Université de Franche-Comté 324.] Paris: Les Belles Lettres, 1986.

Knight, David. "The Growth of European Scientific Monograph Publishing before 1850," 23–41, in Meadows, ed., *Development of Science Publishing in Europe*.

Kronick, David A. *A History of Scientific and Technical Periodicals: The Origins and Development of the Scientific and Technical Press, 1665–1790*. 2nd ed. Metuchen, NJ: Scarecrow Press, 1976.

———. *Scientific and Technical Periodicals of the Seventeenth and Eighteenth Centuries: A Guide*. Metuchen, NJ: Scarecrow Press, 1991.

Kuhn, Thomas S. "Mathematical versus Experimental Traditions in the Development of Physical Science," *Journal of Interdisciplinary History* 7(1976): 1–31. [Reprinted, 31–65 in Thomas S. Kuhn, *The Essential Tension: Selected Studies in Scientific Tradition and Change* (Chicago: University of Chicago Press, 1977).]

Laurent, Gouluen. "Classification," 457–464, in Blay and Halleux, eds., *La Science Classique*.

Lawrence, George H. M., ed. *Adanson: The Bicentennial of Michel Adanson's « Familles des plantes. »* 2 vols. Pittsburgh: The Hunt Botanical Library, 1963–1964.

Lenoir, Timothy, ed. *Inscribing Science: Scientific Texts and the Materiality of Communication.* Stanford, CA: Stanford University Press, 1998.

———. *Instituting Science: The Cultural Production of Scientific Disciplines.* Stanford, CA: Stanford University Press, 1997.

Lovejoy, Arthur O. *The Great Chain of Being: A Study of the History of an Idea.* Cambridge, MA.: Harvard University Press, 1964. [Original edition, 1936.]

Manten, A. A. "Development of European Scientific Journalism Publishing before 1850," 1–22, in Meadows, ed., *Development of Scientific Publishing.*

Meadows, A. J., ed. *Development of Scientific Publishing in Europe.* Amsterdam: Elsevier Science Publishers, 1980.

———. *Communicating Research.* San Diego, CA: Academic Press, 1998.

———. *Communication in Science.* London: Butterworths, 1974.

McClellan, James E. III. "The Académie Royale des Sciences, 1699–1793: A Statistical Portrait." *ISIS* 72 (1981): 541–567.

———. "Académie des Sciences," 1–2, in *Reader's Guide to the History of Science,* Arne Hessenbruch, ed.

———. "Les *Mémoires* de l'Académie Royale des Sciences, 1699–1790. Bilan Public et Processus Privés," in Brian and Demeulenaere-Doyère, eds., *Actes du Colloque.*

———. "The *Mémoires* of the Académie Royale des Sciences, 1699–1790: A Statistical Overview," 7–36, in Halleux et al., *Les publications de l'Académie,* vol. 2.

———. *Science Reorganized: Scientific Societies in the Eighteenth Century.* New York: Columbia University Press, 1985.

———. "Scientific Institutions and the Organization of Science," 99–120, in Roy Porter, ed., *The Cambridge History of Science.*

———. "Scientific Journals," 43–47 in Alan Charles Kors, ed., *Oxford Encyclopedia of the Enlightenment,* vol. 4. New York and Oxford: Oxford University Press, 2002.

———. "The Scientific Press in Transition: Rozier's Journal and the Scientific Societies." *Annals of Science* 36 (1979): 425–449.

Minard, Philippe. "Agitation in the Work Force," 107–123, in Darnton and Roche, eds., *Revolution in Print.*

Moran, Gordon. *Silencing Scientists and Scholars in Other Fields: Power, Paradigm Controls, Peer Review, and Scholarly Communication.* Greenwich, Connecticut and London: Ablex Publishing Corporation, 1998.

Mulkay, Michael J. "Norms and Ideology in Science." *Social Sciences Information,* 15 (1976): 637–656.

Nicolas, Jean-Paul. "Adanson, the Man," 1–121, in Lawrence, ed., *Adanson,* vol. 1.

Paul, Charles B. *Science and Immortality: The Éloges of the Paris Academy of Sciences,*

1699–1799. Berkeley, Los Angeles and London: University of California Press, 1980.

Perkins, Jean A. "Censorship and the Académie des sciences: a case study of Bonnet's *Considérations sur les corps organisés*." *Studies on Voltaire and the Eighteenth Century* 199 (1981): 251–262.

Pinault-Sørensen, Madeleine. "Les dessinateurs de l'Académie des sciences. Status et activités," in Brian and Christiane Demeulenaere-Doyère, eds. *Actes du Colloque, «Règlement, usages et science dans la France de l'Absolutisme.»*

Poirier, Jean-Pierre. *Histoire des femmes de science en France: Du Moyen Age à la Révolution*. Paris: Pygmaliion/Gérard Watelet, 2002.

Porter, Roy, ed. *The Cambridge History of Science, Volume 4: The Eighteenth Century*. Cambridge: Cambridge University Press, 2003.

Price, Derek J. de Solla. *Big Science, Little Science . . . and Beyond*. New York: Columbia University Press, 1986.

Roche, Daniel. "Censorship and the Publishing Industry," 3–27, in Darnton and Roche, eds., *Revolution in Print*.

Roger, Jacques. *The Life Sciences in Eighteenth-Century French Thought*, Keith R. Benson, ed., Robert Ellrich, trans. Stanford, CA: Stanford University Press, 1997.

[Royal Society of London.] *Diplomata et statuta regalis societatis Londini pro scientis naturali promovenda. Iussu Praesidis et Concilii edita*. Londoni: Typic Sam. Richardsoni, MDCCLII.

Rudwick, Martin J. S. "Charles Darwin in London: The Integration of Public and Private Science," *ISIS* 73 (1982): 186–206.

Schaffer, Simon. "The Leviathan of Parsonstown: Literary Technology and Scientific Representation," 182–222, in Lenoir, ed., *Inscribing Science*.

Schiebinger, Londa. "Exotic Abortifacients: The Global Politics of Plants in the 18th Century." *Endeavour* 24 (2000), 117–121.

———. *Nature's Body: Gender in the Making of Modern Science*. Boston: Beacon Press, 1993.

Sgard, Jean, ed. *Dictionnaire des journaux, 1600–1789*, 2 vols. Paris, 1991.

Sgard, Jean, ed., in collaboration with Michel Gilot and Françoise Weil. *Dictionnaire des Journalistes, 1600–1789*. Grenoble, 1976.

Shapin, Steven. *The Scientific Revolution*. Chicago and London: University of Chicago Press, 1996.

Shapin, Steven and Simon Schaffer. *Leviathan and the Air Pump: Hobbes, Boyle, and the Experimental Life*. Princeton: Princeton University Press, 1985.

Sharlin, Harold I. *The Convergent Century: The Unification of Science in the Nineteenth Century*. London and New York: Abelard-Schuman, 1966.

Smeaton, W. A. "*L'Avant-Coureur*. The Journal in which Some of Lavoisier's Earliest Research was Reported." *Annals of Science* 13 (1957): 219–224.

Smith, Pamela. *The Business of Alchemy: Science and Culture in the Holy Roman Empire*. Princeton, NJ: Princeton University Press, 1994.

Speck, Bruce W. *Publication Peer Review: An Annotated Bibliography.* [Bibliographies and Indexes in Mass Media and Communications, Number 7.] Westport, Connecticut and London: Greenwood Press, 1993.

Stewart, Larry. *The Rise of Public Science: Rhetoric, Technology, and Natural Philosophy in Newtonian Britain, 1660–1750.* Cambridge: Cambridge University Press, 1992.

Stroup, Alice. *A Company of Scientists: Botany, Patronage, and Community at the Seventeenth-Century Parisian Royal Academy of Sciences.* Berkeley: University of California Press, 1990.

———. "L'Académie royale des sciences et la censure: l'affaire Duclos, 1677–1685," in Brian and Demeulenaere-Doyère, eds. *Actes du Colloque, «Règlement, usages et science dans la France de l'Absolutisme.»*

———. *Royal Funding of the Parisian Académie Royale des Sciences During the 1690s.* [*Transactions* of the American Philosophical Society, vol. 77, part 4, 1987.] Philadelphia: The American Philosophical Society, 1987.

Sturdy, David J. *Science and Social Status: The Members of the Académie des Sciences, 1666–1750.* Bury St. Edmunds, UK: The Boydell Press, 1995.

Thornton, John L. and R. I. J. Tully, *Scientific Books, Libraries and Collectors.* 3rd ed. London: The Library Association, 1971. Supplement. London, 1978.

Tits-Dieuaide, Marie-Jeanne. "Les Savants, la société et l'état: À propos du «renouvellement» de l'Académie royale des sciences (1699)." *Journal des Savants* (January–June, 1998): 79–114.

———. "Une institution sans statuts: l'Académie royale des sciences de 1666 à 1699," 1–13, in Brian and Demeulenaere-Doyère, eds., *Histoire et mémoire.*

Tort, Patrick. *L'ordre et les monstres : le débat sur l'origine des déviations anatomiques au XVIIIᵉ siècle.* Paris: Le Sycomore [1980].

Vartanian, Aram. "Trembley's Polyp, La Mettrie, and Eighteenth-Century Materialism," *Journal of the History of Ideas* 11 (1950), 259–286.

Velut, Christine. *La Rose et l'orchidée: Les usages sociaux et symboliques des fleurs à Paris au XVIIIᵉ siècle.* [Paris:] Découvrir, 1993.

Vittu, Jean-Pierre. "Périodiques," 140–148, in Michel Blay and Robert Halleux, eds., *La Science Classique.*

Index

Trudaine de Montigny, Jean-Charles-
    Philibert, 22, 101, 119, 120n16
Two-fluid theory of electricity, 53, 54

U

Universities, 94

V

Vaccination with cowpox, 69

Vallière, Joseph-Florent Marquis de, 44
Vandermonde, Alexandre, 119
Variolation against smallpox, 69–71
Vice-president of Académie Royale des
    Sciences, 19, 21–23, 102

W

Waitz, 34
Warnier, 84
Withdrawn papers, 34, 39n39

www.ingramcontent.com/pod-product-compliance
Lightning Source LLC
Chambersburg PA
CBHW061753260326
41914CB00006B/1096